Mathematics Education Library

The *Mathematics Education Library* presents significant new ideas and developments of major importance to those working in the field of mathematics education. The series aims to provide a fundamental resource for educators, researchers and students, and to reflect the variety of research, practice and theory concerns within this discipline. It deals with philosophical, didactical and methodological subjects, as well as with new pedagogical programs in mathematics education.

An understanding of mathematics and of mathematics education is fundamental in our increasingly complex societies where mathematics is expected to be taught at all levels and to students of all ages and abilities. The books published in the *Mathematics Educational Library* reflect this broad scope and are intended to be of universal interest to researchers and teachers in the field of mathematics education.

Book proposals for this series may be submitted to the Publishing Editor: *Melissa James*;

e-mail: Melissa.James@springer.com

More information about this series at http://www.springer.com/series/6276

Alan Bishop • Hazel Tan • Tasos N Barkatsas
Editors

Diversity in Mathematics Education

Towards Inclusive Practices

Springer

Editors
Alan Bishop
Faculty of Education
Monash University
Clayton
Victoria
Australia

Tasos N Barkatsas
School of Education
RMIT University
Bundoora
Victoria
Australia

Hazel Tan
Faculty of Education
Monash University
Clayton
Victoria
Australia

ISSN 0924-4921
ISSN 2214-983X (electronic)
ISBN 978-3-319-05977-8
ISBN 978-3-319-05978-5 (eBook)
DOI 10.1007/978-3-319-05978-5
Springer Cham Heidelberg New York Dordrecht London

Library of Congress Control Number: 2014945341

Printed on acid-free paper

Springer is part of Springer Science +Business Media (www.springer.com)

Introduction to the Book

This book presents a research focus on diversity and inclusivity in mathematics education. The challenge of diversity, largely in terms of student profiles or contextual features, is endemic in western education, and is often argued to require differentiation as a response. This can be seen in the presence of different curricula, text materials, task structures or pedagogies. These however merely emphasise the differentiation status quo, and have had limited success in past practice with changing that situation. For example, huge differences in achievement still exist both within and between countries, states, schools and students. If we in mathematics education seek to challenge that status quo, more research must be focussed not just on diversity but also on the inclusivity of practices in mathematics education.

The book is written by a group of experienced collaborating researchers who share this focus. We met for a writing weekend in early 2013 to explore the intersections between our research projects. In the intervening period, the group met to discuss the structure of the book, the focus of individual chapters, the extent and scope of ideas included, and the time-scale. The book is written for researchers, research students, teachers and in-service professionals who recognise both the challenges as well as the opportunities of creating and evaluating new inclusive approaches to curriculum and pedagogy—ones that take for granted the positive values of diversity. Several chapters report new research in this direction. The overall approach in the book follows the educational approach which we are exploring and in some sense 'promoting'—namely the search for inclusivity within diversity.

The authors are part of, or have visited with, the mathematics education staff of the Faculty of Education at Monash University, in Melbourne, Australia. The chapters all focus on the ideas of growth and development in both research and practice. We do not argue that previous research in this area is irrelevant. Indeed most of the authors have been involved in research which assumed that diversity in practice was both necessary and sufficient. We now recognise however that the current need is for new inclusive approaches which build on that previous research. For example, as societies become more culturally diverse through increasing worldwide immigration, so educational practices need to change to include rather than exclude any students who do not 'fit' the mainstream mould. Additionally, the studies presented were set in different contexts, including Australia, China, the United States and Singapore.

The book builds on previous research on educational differentiation, and is presented in three distinct but related sections which cover a comprehensive range of areas as well as give the book some significant foci. Each section is followed by a chapter reflecting on the prior chapters. The sections are as follows:

Section 1: Surveying the Territory

This section contains six chapters with the first by Peter Sullivan setting the challenge of writing this book and establishing our own inclusivity within our diversity. The chapters all focus on methodological issues and present data from recently completed studies which explore the diversity in large scale data-sets and national policies to infer inclusivity. The section is critiqued by M.A. (Ken) Clements who for many years was a staff member in the Faculty, and who was instrumental in establishing the strong research focus on mathematics education in the Faculty.

Section 2: Interrogating the Boundaries

The three chapters in this section scrutinise the latent aspects of mathematics education in relation to diversity. To some extent this section reflects on the research approaches and theoretical assumptions presented by the chapters in the first section. The critiquing author in this section is Konrad Krainer who has been an occasional visitor to the Faculty and who himself has a strong research interest in the issues of diversity and inclusivity.

Section 3: From Diversity to Inclusive Practices

The three chapters in this section delve into classroom practices and explore some of the challenges and practical opportunities for inclusivity. In some sense the chapters in this section 'close the loop' by referencing the challenges presented by Peter Sullivan in the first chapter of the book. However, in no sense are the chapters in this section presenting 'answers' to those challenges, as might be implied by the heading of this section. What is significant is that in this section the chapters start from considerations of issues of diversity in practice, as an alternative approach to the studies in the previous two sections. Assisting with this analysis is the final critiquing chapter by Laurinda Brown. With a wealth of editorial experience with the outstanding journal *'For the Learning of Mathematics'*, Laurinda is well equipped to offer readers a final (for now) reflection on this highly diverse and complex research area.

As well as acknowledging the contributions of the commenting authors, the editors would like to acknowledge the help given by students, teachers and schools in our various research projects. We also wish to thank Wendy May and Nike Prince for their help with the preparation of the manuscript. Finally, we thank the editorial staff of Springer and the editorial board of the Mathematics Education Library for their constructive advice and feedback about the book.

The Editors

Contents

Part II Interrogating the Boundaries

Part III Towards Inclusive Practices

Contributors

Annica Andersson Faculty of Mathematics and Science Education, Stockholm University, Stockholm, Sweden

Mike Askew Faculty of Education Monash University, Clayton, Australia

Anastasios (Tasos) Barkatsas School of Education Royal Melbourne Institute of Technology (RMIT) University, Melbourne, Australia

Alan J. Bishop Faculty of Education Monash University, Clayton, Australia

Jennifer Bleazby Faculty of Education Monash University, Clayton, Australia

Laurinda Brown Graduate School of Education, University of Bristol, Bristol, UK

M. A. (Ken) Clements Illinois State University, Normal, USA

Helen Forgasz Faculty of Education Monash University, Clayton, Australia

Sarah Hopkins Faculty of Education Monash University, Berwick, Australia

Penelope Kalogeropoulos Faculty of Education Monash University, Clayton, Australia

Konrad Krainer School of Education, Alpen-Adria-Universität Klagenfurt, Klagenfurt am Wörthersee, Austria

Gilah Leder Faculty of Education Monash University, Clayton, Australia

Sarah Lubienski Department of Curriculum and Instruction, University of Illinois, Urbana-Champaign, USA

Debra Panizzon Faculty of Education Monash University, Clayton, Australia

Carly Sawatzki Faculty of Education Monash University, Clayton, Australia

Wee Tiong Seah Faculty of Education Monash University, Clayton, Australia

Peter Sullivan Faculty of Education Monash University, Clayton, Australia

Hazel Tan Faculty of Education Monash University, Clayton, Australia

Celéste de Villiers Graduate School of Education, University of Western Australia, Crawley, Australia

Author Biographies

Andersson Annica is a Senior Lecturer of Mathematics Education at the Faculty of Mathematics and Science Education, Stockholm University and a Guest Senior Lecturer at the Department of Mathematics Education at Linnaeus University in Växjö, Sweden. She has a long experience of mathematics teacher education and research supervisor. Annica's research interests concern two different strands. Her thesis was focussed on students' narratives and identities, in a research setting inspired by *critical mathematics education* (CME). She now continues this work in schools where teaching with ICT is the main pedagogy. The second strand concerns individuals' values and valuing related to mathematics education. Annica leads the Swedish research group on values within the 'Third Wave Project', an international research consortium which coordinates research studies into the harnessing of values in school mathematics education. Annica is the Swedish coordinator of IOWME (International Organization of Women and Mathematics Education) and a board member of SMDF (the Swedish Society for Research in Mathematics Education).

Mike Askew was until recently Monash Foundation Chair Professor of Primary Education and is now an Adjunct Professor there. Previously, he was Chair Professor of Mathematics Education at King's College London. He has directed many research projects including the influential 'Effective Teachers of Numeracy in Primary Schools', which was drawn upon by the Australian Federal Government in developing policy on numeracy teaching. He was deputy director of the five-year Leverhulme Numeracy Research Programme, examining teaching, learning and progression in numeracy to students from age 5 to 11. The findings from such research have influenced policy in England and internationally.

Anastasios (Tasos) N. Barkatsas is a Senior Lecturer, Master of Teaching Practice (Secondary) Program Manager and Graduate Diploma in Education (Primary) Program Director, at the School of Education, RMIT University, Australia. He has been a Senior Research Fellow at the Centre for Educational Research of the Hellenic Republic (Greece) and a Senior Research Fellow at the Pedagogical Institute (now the Institute for Educational Policy) of the Hellenic Republic. He has also been a Visiting Scholar and Adjunct Professor at the National University of Athens. Dr Barkatsas was the Statistical Advisor for Higher Degrees by Research at the Faculty of Education, Monash University, 2011–2013 and he is interested in applications of multivariate statistics in education. He is currently the Chief Quantitative Analyst for the WIFI study of the 'Third Wave Project', an international research consortium which coordinates research studies into the harnessing of values in mathematics education. Dr Barkatsas has also been Head of Mathematics, Science and IT, Head of Curriculum, Deputy Principal and Acting Principal in various secondary Colleges in Melbourne. Dr Barkatsas has published numerous international journal articles, books, book chapters and refereed conference papers. He is an Editorial Board member and Special Quantitative Research and Statistical Modelling Advisor, Journal of *International Research in Early Childhood Education* (IRECE) and an Editorial Board Member, *GAZI Journal of Education* (GAZIJE).

Alan J. Bishop was Professor of Education and Associate Dean at Monash University from 1992 to 2002 after spending the earlier part of his life in the UK. He edited (from 1978 to 1990) the international research journal *Educational Studies in Mathematics*, published by Kluwer (now Springer), and he is Managing Editor of the research book series *Mathematics Education Library*, also published by Springer (1980–present). He was the Chief Editor of two *International Handbooks of Mathematics Education* (1996 and 2002) published by Springer, and joint Editor of the *Third International Handbook of Mathematics Education* (2012), also published by Springer. He was the sole Editor of the *Handbook on Mathematics Education*, published by Routledge (2003).

Jennifer Bleazby is a lecturer in the Faculty of Education, Monash University. She is a former secondary school teacher. In 2008, she completed a PhD in philosophy of education. Her research interests include Philosophy in Schools; the philosophy and educational theory of John Dewey; feminist philosophy of education; epistemology and education, including the role of emotions and imagination in knowing and learning; and ethics and education. She is the author of *Social Reconstruction Learning: Dualism, Dewey and Philosophy in Schools*, published by Routledge in 2013.

Laurinda Brown is Senior Lecturer at the University of Bristol, Graduate School of Education in the UK. Her research has centered around how to make mathematics accessible to all students and how to support the development of prospective teachers of mathematics. She has a particular love of editing, having been the editor of *For the Learning of Mathematics, Mathematics Teaching and Special Issues of JMTE, ESM and currently ZDM* (due 2015). She especially enjoys researching, writing and editing collaboratively. Alongside her interests in mathematics education, membership of both the Narrative Inquiry Centre (NIC) and Bristol Collaborative Writing Group (BCWG) at the University of Bristol has led to development of qualitative methodologies such as narrative interviewing and coobservation in her mathematics education research.

McKenzie A (Ken) Clements has been Professor in the Department of Mathematics at Illinois State University since 2005. Ken taught in schools for 14 years, and in five universities—Monash University, Deakin University, the University of Newcastle

(Australia), Universiti Brunei Darussalam, and Illinois State University. Overall, he has authored, co-authored or co-edited 27 books including three international handbooks of mathematics education. With his wife, Nerida Ellerton, he has recently written *Rewriting the History of School Mathematics in North America 1607–1861* (Springer 2012) and *Abraham Lincoln's Cyphering Book and Ten Other Extraordinary Cyphering Books* (Springer 2014). In addition, Ken has had more than 200 articles published in refereed journals or as book chapters. He is honorary life member of both Mathematics Education Research Group of Australasia (MERGA) and the Mathematical Association of Victoria.

Celéste de Villiers is a Doctoral student in the Graduate School of Education, University of Western Australia. She has extensive experience teaching at various levels, including early childhood, primary and tertiary (pre-service teacher) levels. She has a longstanding interest in learning difficulties and is currently the learning enrichment co-ordinator at one of Australia's largest private schools. Most recently, she presented a paper at the conference for the Mathematics Education Research Group of Australasia (2013). Celéste holds the degrees of M.Ed., Grad.Cert.Ed., Dip.Ed., and H.Dip.Ed.

Helen J. Forgasz is a professor of education in the Faculty of Education, Monash University. Her research interests include gender and equity issues in mathematics education, the affective domain, learning settings, and numeracy across the curriculum. Helen has published widely in scholarly and professional journals, and has co-edited several monographs on gender issues and mathematics learning. She serves on the editorial boards of high quality journals and book series and continues to be an active member of mathematics education research associations nationally and internationally.

Sarah Hopkins is Senior Lecturer in the Faculty of Education at Monash University. After teaching secondary school mathematics in a variety of settings, including schools located within juvenile justice centers, Sarah now teaches primary and secondary pre-service teachers. Her research activities focus on mathematics learning difficulties, assessments of teaching quality and effectiveness and applications of Rasch measurement. She was awarded a research grant from the Spencer Foundation (Chicago University) to investigate difficulties with simple addition. Most recently she has written a chapter titled 'Understanding and supporting competency in numeracy' for the widely used textbook *Inclusion in Action (3rd Edition)* edited by P. Foreman (2014). Sarah holds the degrees of PhD., M.Ed.St., and B.Ed.

Penny Kalogeropoulos Penelope is a current PhD student at Monash University. Her research concerns students' and teachers' values associated with (dis) engagement in mathematics education. Included in Chapter 10 of this book is a discussion of her (in progress) research data and analysis. As a lead teacher she has worked at Wooranna Park Primary school in Melbourne for 9 years. She focusses on implementing innovative and creative mathematics teaching and learning strategies. Her passion is developing mathematics learning that ignites excitement in children (especially for her own two young boys).

Konrad Krainer is full professor and Director of the School of Education at the Alpen-Adria-University Klagenfurt in Austria. He worked several years as a mathematics teacher and wrote his Doctoral dissertation in the field of geometry education. His recent research focusses on mathematics teacher education, school development and educational system development. He is co-editor of several books (e.g., one volume of the *International Handbook of Mathematics Teacher Education*) and the leader of the nation-wide IMST (Innovations Make Schools Top) project. He was Associate Editor of *Journal of Mathematics Teacher Education*, founding and board member of the European Society for Research in Mathematics Education, and is member of the Education Committee of the European Mathematical Society. He gave several invited lectures and took part in or chaired Plenary panels at international Conferences (e.g. ICM, ICME and PME). He is the Chair of the International Programme Committee for the Ninth Congress of European Research in Mathematics Education (CERME 9, Prague 2015).

Gilah Leder is an Adjunct Professor at Monash University and Professor Emerita at La Trobe University. Her research has focussed particularly on gender issues in mathematics education, on exceptionality—predominantly high achievement, and on the interaction between learning and assessment. She has published widely in each of these areas and continues to serve on the Editorial Board of national and international mathematics education and general education journals. Gilah is a Past President of the Mathematics Research Group of Australasia (MERGA) and of the International Group for the Psychology of Mathematics Education (PME), a Fellow of the Academy of the Social Sciences in Australia, the recipient of the Felix Klein medal awarded for outstanding lifetime achievements in mathematics education research and development and of the 2013 MERGA Career Research Medal.

Sarah Lubienski is a professor of mathematics education and an Associate Dean in the Graduate College at the University of Illinois, Urbana-Champaign. Her research focusses on inequities in students' mathematics outcomes and the policies and practices that shape those outcomes. She examines social class, race/ethnicity, and gender in large-scale studies using national datasets, as well as in smaller, classroom-based studies. Dr Lubienski has chaired the National Assessment of Educational Progress Research Special Interest Group of the American Educational Research Association (AERA), and is a member of AERA's Grants Governing Board. Dr Lubienski is Principal Investigator and Director of the University of Illinois Postdoctoral Research Training Program in Mathematics Education, funded by the US Institute of Education Sciences. Her work has also been funded by the National Center of Education Statistics, the National Science Foundation and the Fulbright Program.

Debra Panizzon is an Associate Professor of Science Education at Monash University having held previous academic positions at Flinders University and the University of New England. Prior to commencement in academia she taught junior science in secondary schools along with senior Biology and Physics. Debra is an experienced science education academic having worked with both primary and secondary pre-service teachers. Her research interests lie in the areas of cognition, student acquisition of scientific concepts, assessment and rural and regional education given extensive experience living and working in rural New South Wales, Australia. Importantly, much of her research has emerged from partnerships with science and mathematics teachers ensuring that theory and practice are inextricably linked.

Carly Sawatzki is a teacher-educator in the Faculty of education, Monash University. She has curriculum expertise in Economics and Business, Business Management and Psychology. Carly is currently undertaking a PhD exploring the role of social and mathematical understandings in children's financial problem-solving and decision-making. She is particularly interested in the ways and means by which family values, attitudes and expectations regarding money seem to influence the way students approach mathematical problems situated in financial contexts. The study involves developing, trialling, studying and refining unique financial dilemmas and associated pedagogies to enhance financial literacy teaching and learning.

Wee Tiong Seah is Senior Lecturer of Mathematics Education at the Faculty of Education, Monash University, Australia. He is a teacher educator with international and national recognition for his research on the use of the volitional variable of values to activate students' cognitive skills and affective dispositions relating to mathematics learning and teaching. Wee Tiong currently leads the 'Third Wave Project', an international research consortium which coordinates research studies into the harnessing of values in school mathematics education. He was invited to co-edit a special issue of the '*ZDM: The International Journal on Mathematics Education*' in 2012, which highlighted the theme of values in mathematics education. Wee Tiong is also featured in *Encyclopedia of Mathematics Education*, Springer, 2014 for one of his studies investigating how immigrant teachers of mathematics negotiate about value differences and conflicts in the mathematics classroom. Wee Tiong is a member of editorial boards of several journals, including the *International Journal of Science and Mathematics Education.*

Peter Sullivan is Professor of Science, Mathematics and Technology at Monash University. His main professional achievements are in the field of research. His recent research includes four Australian Research Council (ARC) funded projects. He was a member of the Australian Research Council College of Experts for Social Behavioural and Economic Sciences for 4 years. He is an author of the popular teacher resource *Open-ended maths activities: Using good questions to enhance learning* that is published in the US as *Good questions for math teaching*. Until recently he was chief editor of the *Journal of Mathematics Teacher Education*, is immediate past president of the Australian Association of Mathematics Teachers, was the author of the Shape paper that outline the principles for the development of the Australian Curriculum in Mathematics, and was the author of the 2011 Australian Education Review on research informed strategies for teaching mathematics.

Hazel Tan is a lecturer in the Faculty of Education at Monash University. She has recently completed her PhD investigating factors influencing senior secondary mathematics students' interactions with advanced calculators. She has taught senior secondary mathematics for many years, and was the Head of Mathematics Department in a Singaporean school. Hazel has also worked in the Educational Technology Division of Singapore's Ministry of Education, spearheading the pedagogical use of technologies in education. Her research interests are in the area of teaching and learning secondary mathematics with technologies and related teacher education, comparative education and gender issues in mathematics education.

Part I
Surveying the Territory

Chapter 1
The Challenge of Reporting Research to Inform the Creation of Inclusive Mathematics Learning Environments

Peter Sullivan

Introduction

This book draws on selected research of its chapter authors to describe various threats to inclusive education. The focus is on mathematics learning since that domain so often serves to exclude some students from progression. While the major threats to inclusiveness are well known, various chapters articulate the threats in ways that illustrate and elaborate the processes through which the threats restrict students' opportunity to learn. More importantly, the chapters raise the issue of how such threats can be addressed. Of course, it is much more difficult to redress student differences than it is to identify them. It is also possible that steps that education providers take to redress differences can sometimes exacerbate the exclusion of some students. There are even important fundamental issues that are still debated. One such issue for this book is whether the goal of any recommendations for change is to improve the education of all students, without addressing the differences, or to find ways to reduce the differences between groups of students. The critical role of education in preparing for future economic growth is recognised across the world. In Australia, for example, at a time of significant attention to the funding of schools and to addressing inequities in education opportunities, it is incumbent on researchers to go beyond describing factors contributing to differences in opportunities and learning and to propose methods of redressing the differences. This introductory chapter elaborates some of the challenges of reporting research to inform schools and teachers on ways of addressing differences.

The chapter starts by considering the nature of the mathematics that all students should learn since sometimes decisions on that mathematics contribute to student exclusion. Some of the challenges in using research to offer advice to teachers are then discussed using a particular context. An attempt to convert research findings on pedagogy into suggestions for teachers is presented, and some challenges with

P. Sullivan (✉)
Faculty of Education, Monash University, Clayton, Australia
e-mail: peter.sullivan@monash.edu

A. Bishop et al. (eds.), *Diversity in Mathematics Education*,
Mathematics Education Library, DOI 10.1007/978-3-319-05978-5_1,

implementing such advice are then discussed. The theme is that drawing practical advice from research is complex but it is also an imperative for researchers if they are to contribute positively to solutions.

As with other chapters in the book, the research context is the Australian school system and policy environment generally. While many of the chapters outline various aspects of the system to indicate ways in which they are idiosyncratic, the issues addressed in this chapter are applicable in many countries and cultures worldwide.

A Perspective on the Numeracy and Mathematics that Students Might Learn

At a recent meeting with academics who specialise in the teaching of English language, there was a discussion of whether students from low socioeconomic backgrounds should study Hamlet. The consensus was that these are the very students who most need to study such key elements of English literature. The English educators argued that while functional literacy is important so is exposure to literature since it connects those students with culture and tradition. It is interesting to consider the parallel argument in the case of mathematics which is that all students need exposure to both mathematical literacy, termed numeracy in the following, and to mathematics. The point being that whether a researcher is exploring either or both of these options, the focus needs to be explicit.

It seems that the common tendency in schools to offer some students a limited curriculum focusing solely on practical mathematics or low-level routines restricts their study and career opportunities and also contributes to their alienation from school. It is noted that the term numeracy is sometimes used to describe this restricted curriculum, although in other places, such as in systemic assessments, the term is taken to be synonymous with mathematics.

There is a general agreement among policy makers, curriculum planners, school administrators, and business and industry leaders that mathematics is an important element of the school curriculum because of the potential economic benefits to the broader society. Indeed, the importance of numeracy and mathematics is implicitly accepted by governments in the emphasis they place on monitoring school improvement in mathematics in the public reporting of student results. Yet there is still an on-going debate among mathematics educators on what aspects of mathematics are important, and which aspects of mathematics are most needed by school leavers.

On one hand, commentators argue for the need to intertwine conventional discipline-based learning with practical perspectives, whilst on the other hand there are those who argue for a focus on the mathematics skills needed for higher study.

In terms of the former, it is argued that schools are confronting serious challenges from disengaged students. In their report on the national Middle Years Research and Development Project, Russell et al. (2013), for example, made recommendations for reform associated with school leadership and systematic school improvement, especially emphasising the need for more interesting, functionally relevant classroom tasks can enhance engagement in learning. Klein et al. (1998) had previously argued that the mathematics taught in schools should focus on practical aspects preparing school leavers for employment and for their everyday needs as citizens.

Yet, in terms of the latter, there is a decline in the number of students completing later year university level mathematics studies, and the argument is that this threatens Australia's future international competitiveness and capacity for innovation. It is often suggested that the solution is for more mathematical rigour at secondary level, as preparation for more advanced learning in mathematics. Rubenstein (2009), for example, offers a compelling description of the economic challenges Australia is facing due to the decline in mathematics enrolments in latter year university mathematics studies.

Unfortunately, these arguments are presented as though teachers must adopt one perspective or the other. Rather than seeking to focus on one or the other side of this debate, it is preferable that curricula encompass both perspectives, with appropriate variations according to the learners' backgrounds. In other words, all students should experience not only practical uses of mathematics but also the more formal aspects that lay the foundation for later mathematics and related study. The key is to identify the relative emphases, and the foci within each perspective, relevant to the learners.

Ernest (2010), in addressing these two foci for mathematics teaching, described the goals of the practical/numeracy perspective as being that students learn the mathematics adequate for general employment and functioning in society, drawing on the mathematics used by various professional and industry groups. He included in this perspective the types of calculations one does as part of everyday living including best buy comparisons, time management, budgeting, planning home maintenance projects, choosing routes to travel, interpreting data in the newspapers, and so on.

Ernest also described the specialised perspective as the mathematical understanding which forms the basis of university studies in science, technology, and engineering. He argued that this includes an ability to pose and solve problems, appreciate the contribution of mathematics to culture, the nature of reasoning, and intuitive appreciation of mathematical ideas such as

> …pattern, symmetry, structure, proof, paradox, recursion, randomness, chaos, and infinity. (Ernest 2010, p. 24)

The importance of both perspectives is evident in the discussions which informed the development of the Australian Curriculum: Mathematics. For example, the Australian Curriculum Assessment and Reporting Agency (ACARA) (2010) Mathematics Shape Paper listed the aims of emphasising the practical aspects of the mathematics curriculum as being:

> …to educate students to be active, thinking citizens, interpreting the world mathematically, and using mathematics to help form their predictions and decisions about personal and financial priorities. (ACARA 2010, p. 5)

It also described the specialised aspects as including that:

> …mathematics has its own value and beauty and it is intended that students will appreciate the elegance and power of mathematical thinking, (and) experience mathematics as enjoyable. (ACARA 2010, p. 5)

In other words, ACARA requires the national curriculum in mathematics to incorporate both perspectives. The key issue rests in determining their relative emphases. Ernest (2010) argued that, while it is important that students be introduced to aspects of specialised mathematical knowledge, the emphasis in the school curriculum for the compulsory years should be on practical mathematics. Another perspective on this point was made by Ainley et al. (2008) who noted that whilst fewer than 0.5 % of university graduates specialise in mathematics, and that only around 40 % of graduates are professional users of mathematics, a full 100 % of school students need practical mathematics to prepare them for work as well as for personal and social decision-making.

A possible conclusion is that the priority in the compulsory years should be mathematics of the practical perspective. While the education of the future professional mathematicians is not to be ignored, the needs of most school students are broader. The term *numeracy* is commonly taken by Australian policy makers and school practitioners to incorporate the practical perspective of mathematical learning as the goal for schools and mathematical curricula. The implication is that an emphasis on numeracy should inform curriculum, pedagogy, and assessment in mathematics and even in other disciplines, especially in the compulsory school years. On balance, it seems that the curriculum would be better to address both specialised and practical perspectives concurrently. Unfortunately, it also seems that neither perspective is being done well in Australian schools, and indeed schools in many other countries as well.

The *Third International Mathematics and Science Study*, which aimed to investigate and describe Year 8 mathematics and science teaching across seven countries, makes an interesting contribution to this debate. In the Australian component, 87 teachers, each from a different school volunteered and this cohort provided representative regional and sectoral coverage across all Australian States and Territories. Each teacher in their mathematics class was filmed for one complete lesson. With respect to the Australian lessons, Hollingsworth et al. (2003) found that most exercises and problems used by teachers were low in procedural complexity, that most were repetitions of problems that had been previously completed, that little connection was made to examples of uses of mathematics in the real world, and that the emphasis was on students locating just the one correct answer. They concluded:

> Opportunities for students to appreciate connections between mathematical ideas and to understand the mathematics behind the problems they are working on are rare. (Hollingsworth et al. 2003, p. xxi)

Similarly, Stacey (2010) reported findings from a recent study in which she interviewed over 20 leading educators, curriculum specialists, and teachers on their perspectives on the nature of Australian mathematics teaching. She concluded that the consensus view is that Australian mathematics teaching is generally repetitious, lacking complexity, and rarely involves reasoning.

Such mathematics teaching seems common in other countries as well. For example, Swan (2005), in summarising reports from education authorities in the UK, concluded that much mathematics teaching there consisted of low-level tasks that could be completed by mechanical reproduction of procedures without deep thinking. Swan argued that students of such teachers are mere receivers of information, having little opportunity to actively participate in lessons, are allowed little time to build their own understandings of concepts, and they experience little or no opportunity or encouragement to explain their reasoning.

To elaborate the implications for schooling which supports inclusion, it is common for teachers in the junior and middle secondary years to group students identified as low achieving together and offer them a curriculum that either focuses on remediating skills that have been taught for some years, such as operations with fractions, or routine practical tasks such as measuring. It would be preferable for those students to also engage with age appropriate and meaningful tasks that are aligned with the specialist perspective such as introductory indices, co-ordinate geometry, and probability, for example, rather than repeating low-level tasks that they appear to have not learned previously. Offering a 15-year-old student experiences that they can remember they struggled with when 8 years old is unproductive. It is possible to use the increased maturity and life experiences as a way of engaging such students in both specialised and practical mathematics that is both interesting and useful.

It is evident that the English language educators mentioned above certainly do not see opportunities to experience Hamlet as being exclusive from functional approaches to the learning of language. It seems that a corresponding approach should be possible in mathematics, and teachers would appreciate advice from researchers on how this can be achieved.

Examining Advice About Inclusiveness Drawn from a Particular Context

The thrust of the argument in this chapter is that one of the tasks for education researchers is to go beyond identifying and even describing factors contributing to student exclusion and to elaborate ways that the constraining factors can be addressed. That this seems to be done seldom is partly connected to its complexity. The following explores this issue.

At the time of writing this chapter, I was working at the invitation of the minister of education in the Northern Territory (NT) in Australia on a report to advise on support for the teaching of literacy and numeracy there. The NT has a population of 212,000 out of the 22 million in Australia overall. Approximately 30 % of its population are Indigenous, and by any index most students there are disadvantaged in terms of opportunity and achievement in comparison not only with the rest of Australia but anywhere in the developed world. The minister and the representatives of the Department of Education and Children's Services are serious about inviting recommendations of actions they can take to increase educational opportunities and achievement of students there.

Within the NT, there are four towns that provide many of the resources of regional centres in developed countries. The government has an aspiration that schools serving those centres have educational programs that are equal to the rest of Australia and which can drive future economic growth in the NT, acting as an accessible bridge to Asia. Around 75 % of NT students are from those four centres. The rest of the students attend remote schools, many of which are accessible only by air. In other words, the educational challenges are substantial. The fundamental assumption is that research in mathematics education can make a contribution. This includes advice on actions that policy makers, schools, and teachers can take to improve opportunities and achievement in NT schools. If the findings from the broader body of education literature cannot inform practices in the NT, then it cannot contribute to improving access for marginalised students anywhere. But the challenges of converting research findings to useful and practical advice are substantial.

To illustrate the challenges of offering succinct advice, it is worth examining the type of advice offered on school leadership. For example, Masters (2010) outlines some dispositions and foci for school leadership. It is stressed that the following is not at all intended to criticise the author but merely to illustrate the challenges of converting research knowledge into advice for systems and schools. For example, one aspect of his advice, written as benchmarks against which schools can be evaluated, said:

> The school leadership team has established and is driving a strong improvement agenda for the school, grounded in evidence from research and practice, and couched in terms of improvements in measurable student outcomes. (p. 11)

There are two aspects to the ambiguity of this advice. First, it is hard to imagine what else would be guiding an "improvement agenda". Second, the advice gives little or no indication of what these actions would look like in schools. Perhaps in an attempt to be more specific, Masters (2010) lists the following expectation that:

> Explicit and clear school-wide targets for improvement have been set and communicated, with accompanying timelines. (p. 11)

The same ambiguity is evident in this, with advice on actions also not included. Similarly, there is advice on the use of data that is also somewhat general:

> A high priority is given to the school-wide analysis and discussion of systematically collected data on student outcomes, including academic, attendance and behavioural outcomes. Data analyses consider overall school performance as well as the performances of students from identified priority groups; evidence of improvement/regression over time; performances in comparison with similar schools; and, where possible, measures of growth across the years of school. (p. 11)

There are suggestions on ways that schools should operate in terms of the overall learning environment:

> There is a strong collegial culture of mutual trust and support among teachers and school leaders. The school works to maintain a learning environment that is safe, respectful, tolerant, inclusive and that promotes intellectual rigour. (p. 11)

…and in terms of curriculum programs:

> The school has a coherent, sequenced plan for curriculum delivery that ensures consistent teaching and learning expectations and a clear reference for monitoring learning across the

> year levels. The plan, within which evidence-based teaching practices are embedded, and to which assessment and reporting procedures are aligned, has been developed and refined collaboratively to provide a shared vision for curriculum practice. This plan is shared with parents and caregivers. (p. 11)

…as well as pedagogy:

> The school principal and other school leaders recognise that highly effective teaching practices are the key to improving student learning throughout the school. They take a strong leadership role, encouraging the use of research-based teaching practices in all classrooms to ensure that every student is engaged, challenged and learning successfully. All teachers understand and use effective teaching methods—including explicit instruction—to maximise student learning. (p. 11)

While there is no quibble with any of this (with the exception perhaps of the comment on explicit instruction which might be misinterpreted), the point is that the advice does not inform the type of actions that schools and teachers can take to address the exclusion that many students in the NT experience. It is incumbent on researchers to make the transfer of findings from research into practice more explicit to inform principals and teachers seeking to overcome the barriers that some students are experiencing in their mathematics learning.

Making Specific Recommendations About Mathematics Education Pedagogy

As a way of contributing further to the consideration of the processes of reporting on implications from research to inform teacher and school practices, the following are some recommendations that are intended to offer guidance to teachers of mathematics.

The following unpublished advice was formulated in the same context of offering advice to schools and teachers in the NT on approaches to numeracy and mathematics teaching. The intention is that this extract from the advice relates to pedagogies that can improve access to all students. It is presented to prompt consideration of whether it is possible to offer specific advice drawn from research, and if so whether the following approximates that advice.

The advice is drawn from a review of research-based teaching strategies (Sullivan 2011) that drew on a range of sources such as Good et al. (1983) who synthesised findings on active teaching, from the important summary of meta analyses, Hattie (2009), from other reviews of advice such Clarke and Clarke (2004) who were reporting findings from the Early Numeracy Research Project, and from the Anthony and Walshaw (2009) best evidence synthesis.

The following recommendations about approaches to the teaching of mathematics and numeracy are written in the form of advice to teachers. Nevertheless there are implications in these recommendations for system and school leaders, for leaders of teacher professional learning including those in universities, and for resource developers. The recommendations are as follows:

> Use the Australian Curriculum to identify important ideas that underpin the numeracy concepts you are seeking to teach, communicate to students that these are your intentions for their learning, and explain to them the processes they will use for that learning.

Perhaps, the main benefit of articulating learning intentions is the clarity and focus it provides for the teachers (see, Hattie and Timperley 2007). It is important to note, though, that the teachers' intentions for learning should not reduce the capacity of students to determine their own approaches nor should the intentions set low limits for student achievement. The advice continues:

> Using culturally responsive approaches, build on what the students know, both mathematically and experientially, including creating and connecting students with stories that both contextualise and establish a rationale for the learning.

While all students in the NT should have the same access to the full mathematics and numeracy curriculum as do other Australian students, the approach to teaching that content may be productively adapted to suit the experience of the students. This aligns with what is termed "culturally and contextually responsive" pedagogies (see Australian Association of Mathematics Teachers 2013) and includes approaches that increase opportunities for student decision-making, that give students time for deep learning, and that emphasise respectful communications (Sullivan and Mornane 2014). Generally, the rule "experience before instruction" applies. In terms of experiences for students:

> Engage students in a range of age appropriate connected experiences that allow students opportunities to make decisions about emphases and approaches, and which use a variety of forms of representation.

There are a number of aspects to this. First, classroom experiences are more engaging when they allow students opportunities to build connections between ideas. Second, it is preferable that students have opportunities to make decisions which make it clear to them that they can learn. This is the opposite of experiences in which students are told what to do and which they then implement: this is disabling. Third, the range of experiences and the variety of forms of representation address the diversity of learning styles evident in all classrooms. Fourth, whatever their mathematical background, students are more likely to engage in experiences that they consider purposeful and appropriate to their maturity. This is especially relevant for adolescents.

The advice also emphasised the notion of a classroom community (Sullivan et al. 2009), which incidentally is contrary to many recommended approaches which refer to individualising student programs:

> Create a supportive classroom community in which all students engage with experiences that encourage them to connect ideas together, to describe their approaches to others and to listen when others are explaining their thinking.

Communication is an essential part of learning numeracy, but this creates another barrier for students who are not learning in their first language. Particular teacher actions to address this can include encouraging students to use home language at times (see Boaler and Staples 2008), modelling desired processes while minimising unnecessary talking, the teacher rephrasing student-written solutions on behalf of the students, the use of erasable mini white boards on which students can make private attempts at solutions but which are accessible by the teacher.

Another challenge facing many teachers is the irregular attendance of students. It is therefore productive if experiences are self-contained, meaning they rely as little as possible on prerequisite experiences from prior lessons. This connects directly to the ways that individual differences are addressed:

> Differentiate classroom experiences in ways that preserve the possibility of all students engaging with the classroom community.

Differentiation includes specifically planning to support students who need it, and extending those who are ready within the context of a coherent classroom community. It is possible to plan to support learners experiencing difficulty by changing the form of representation including making the experiences more physical, reducing the number of steps, and lessening the size of quantities involved (see Sullivan et al. 2009). It is also important to plan experiences for those students who are ready for further challenges, at least part of which is communicating expectations for their learning:

> Communicate your expectations that students will engage in experiences which are appropriately challenging for them.

Learning is more robust when students connect ideas together for themselves, and determine their own strategies for solving problems, rather than following instructions they have been given. This occurs when students engage with challenging experiences which require them to plan their approach, especially sequencing more than one step; process multiple pieces of information with an expectation that they make connections between those pieces and see concepts in new ways; choose their own strategies, goals, and level of accessing the task; spend time on the task and record their thinking; be willing to take risks and explain their strategies; and justify their thinking to the teacher and other students.

The expectations communicated to students include aspirations for their behaviour, attendance, participation, and effort as well as achievement. This also includes modelling persistence, affirming persistence when it happens (see Dweck 2000), and recognising failure as a step on the path to powerful learning.

The advice also recognised that the development of fluency with mathematics and numeracy is important:

> Provide opportunities for students to become fluent with their numeracy. This can be done through short everyday practice of mental calculation or number manipulation; and through specific practice, reinforcement and prompting transfer of learnt skills.

Sometimes teachers emphasise fluency to the detriment of other aspects of mathematics and numeracy. The reverse is also true. This point is about reminding teachers that planned, systematic review and practice contribute to fluency in numeracy which in turn contribute to other aspects of numeracy.

Of course, there are other aspects of pedagogy arising from research that are important and which could be considered for inclusion in such a set of recommendations.

Readers can judge whether the above is clear or comprehensive or necessary or sufficient. Even so, it is suspected that teachers would appreciate further advice, drawn from research into practice, on how to implement such recommendations.

Recognising the Challenges in such Advice

To elaborate the need to go further than general lists of advice, the following draws on some findings from Sullivan et al. (2010) who reported on a series of teaching explorations at an Indigenous Community School in a remote region of Western Australia. The project was seeking ways to support the teaching of mathematics in small community-run schools. In the lessons observed, there were many instances that would be judged as outstanding teaching and learning in any school, and certainly demonstrated that students in remote schools can learn as well as their metropolitan counterparts. Yet there were challenges. The following are four examples of these challenges that arose when project teachers were seeking to implement the type of teaching recommendations included in the previous section.

The first challenge is that care needs to be taken when making inferences about the extent of student engagement. In the sequence of activities the class seemed highly engaged. Yet in subsequent interviews, while most students in the class observed were highly fluent with the relevant concepts, there were two students who were not able to respond to even basic questions. Yet those students somehow had been able to mask their lack of understanding which was not noticed by either the teacher or the observer. This emphasises that there is a diversity of achievement within each class, and a diversity of readiness. Ideally, specific actions can be taken to accommodate this diversity which avoid the tendency to include such students by "telling", the net effect of which is to develop in those students a dependence on the teacher.

A second issue relates to the conduct of whole class discussions after rich explorations. The teacher observed often patiently probed student thinking and invited them to explain their reasoning. Yet this was not often successful from a whole-class perspective. One example was the student who explained his strategy for winning a game. He gave an extended explanation and, if one knew what he was trying to say, his explanation was insightful, and illustrated clear conditional thinking and argument. Yet his explanation would not have informed other listeners. There were a number of other instances where an individual gave an excellent explanation that elaborated on the desired type of thinking, but not in a way that would engage the other children. The other students were not interested in such explanations, which may be partly a function of this lack of clarity. The teacher was energetic and committed to this approach, and had worked with the class on her expectations for participation. It is suspected that specific actions are necessary so that this aspect of the approach can realise its potential. One strategy that seemed to work was for the teacher to restate the explanations given by students, and to provide additional diagrammatic support for their explanations. The use of erasable mini whiteboards also can work in that students can display their work just to the teacher and the teacher can explain the work of selected individual to the class which can have a similar effect to the student reporting directly.

A third issue is the common response by students to teacher questions by calling out responses which has the effect of encouraging students to guess. One of our project teachers invited students to whisper answers to the student next to them which moved the students beyond approaches that relied on guessing.

A further issue is the intensity of the interactivity especially since teachers are encouraged to be active and to communicate high expectations. In many of the observations, the students became tired. Noting that these classes are quite small, students are constantly under scrutiny. In the lesson observations, there were mathematically rich and challenging experiences in which the students participated well, even beyond expectations. But it is perhaps unreasonable to expect the students to do this for the full 90 min of each mathematics class. One approach could be for teachers to plan some experiences that are less intensive and less interactive and these could be used to buffer shorter and more intensive parts of the lessons. These less intensive experiences could include competitive games, including card games, or some aspect of physical activity combined with a mathematical experience, or drawing, or storytelling.

The implication of these is that the basic advice is not enough. Advice on pedagogies, for example, needs to be tested as part of the research process and the initial hypotheses adjusted to increase the potential effectiveness of the advice.

Conclusion

The basic argument of this chapter is that it is incumbent on education researchers to be explicit about the implications of their findings. In the case of research that is addressing inclusiveness; one aspect of this advice on implementation is what redressing disadvantage might look like. On one hand, the goal might be to reduce differences between groups of students. On the other hand, the goal might be to improve the outcomes of all students. Certainly the intention is that the effects of factors that might inhibit the learning of particular groups of students such as their parent income, Aboriginality, where they live, where they or their parents were born, whether the home language matches the language of instruction, or their gender would be minimised.

The chapter argued that researchers should go beyond identifying the existence or even the causes of differences to consider what can be done by systems, schools, and teachers to address differences. This includes adopting a clear perspective on the mathematics and numeracy that is to be taught and how they might be taught. It was argued that age appropriate experiences are more likely to enhance the inclusion of marginalised students than merely activities that are matched to the levels achieved by the students on systemic or standardised assessments. It was argued that constructing such advice is complex but researchers and scholars who interpret the research are encouraged to find ways to offer suggestions that can guide practice.

This orientation is relevant at the design stages of research, during the data collection, and in the interpretation of results. Of course research findings ideally make contributions to existing results and to the development of theory. Yet if the goal is to influence practice, and especially to inform approaches to equity, then articulating implications for practice is a consideration. One approach taken by many researchers is to include practitioners in their research teams.

The various chapters in this volume explore this theme more fully. For example, chapters 10, 11, 12 and 13 each specifically connects arguments about inclusion to classroom practice. Chapters 7, 8 and 9 offer philosophical perspectives that can inform the decisions teacher make. There are also chapters that connect large-scale data with specific teacher actions. For example, chapter 2 explores implications from large-scale test data for classroom teachers, chapter 3 articulates ways that isolation acts as a barrier to some students, and chapter 4 examines student preferences for particular tasks and the implications for teachers' selection of tasks. Chapters 5 and 6 explore issues associated with gender stereotypes and the implications that this can have for the education of girls.

References

Ainley, J., Kos, J., & Nicholas, M. (2008). *Participation in science, mathematics and technology in Australian education*. ACER Research Monograph 4.

Anthony, G., & Walshaw, M. (2009). *Effective pedagogy in mathematics. Educational Series-19*. Brussels: International Academy of Education; Geneva: International Bureau of Education.

Australian Association of Mathematics Teachers. (2013). Make it count. http://makeitcount.aamt.edu.au/.

Australian Curriculum Assessment and Reporting Authority (ACARA). (2010). The shape of the Australian Curriculum: Mathematics (p. 4). Found at http://www.acara.edu.au/phase_1_-_the_australian_curriculum.html. Accessed Sept 2013.

Boaler, J., & Staples, M. (2008). Creating mathematical futures through an equitable teaching approach: The case of Railside school. *Teachers College Record, 110*(3), 608–645.

Clarke, D. M., & Clarke, B. A. (2004). Mathematics teaching in Grades K-2: Painting a picture of challenging, supportive, and effective classrooms. In R. N. Rubenstein & G. W. Bright (Eds.), *Perspectives on the teaching of mathematics* (66th Yearbook of the National Council of Teachers of Mathematics, pp. 67–81). Reston: NCTM.

Dweck, C. S. (2000). *Self theories: Their role in motivation, personality, and development*. Philadelphia: Psychology Press.

Ernest, P. (2010). *The social outcomes of learning mathematics: Standard, unintended or visionary? Make it count: What research tells us about effective mathematics teaching and learning* (pp. 21–26). Camberwell: ACER.

Good, T. L., Grouws, D. A., & Ebmeier, H. (1983). *Active mathematics teaching*. New York: Longmans.

Hattie, J. (2009). *Visible learning: A synthesis of over 800 meta analyses relating to achievement*. New York: Routledge.

Hattie, J., & Timperley, H. (2007). The power of feedback. *Review of Educational Research, 77*(1), 81–112.

Hollingsworth, H., Lokan, J., & McCrae, B. (2003). *Teaching mathematics in Australia: Results from the TIMSS video study* (TIMSS Australia Monograph No. 5). Camberwell: Australian Council for Educational Research.

Klein, A. S., Beishuizen, M., & Treffers, A. (1998). The empty number line in Dutch second grades: Realistic versus gradual design. *Journal for Research in Mathematics Education, 29*(4), 443–464.

Masters, G. (2010). Improving educational outcomes in the northern territory. http://www.education.nt.gov.au/__data/assets/pdf_file/0007/19690/ImprovingEducationalOutcomesNT.pdf. Accessed Sept 2013.

Rubenstein, H. (2009). A national strategy for mathematical sciences in Australia. Strategy document written in consultation with the Australian council of Heads of Mathematics Science. http://www.amsi.org.au/pdfs/National_Maths_Strategy.pdf. Accessed Sept 2013.

Russell, V. J., Mackay, T., & Jane, G. (2003). Messages from MYRAD (Middle Years Research and Development): Improving the middle years of Schooling, IARTV.

Stacey, K. (2010). *Mathematics teaching and learning to reach beyond the basics. Make it count: What research tells us about effective mathematics teaching and learning* (pp. 21–26). Camberwell: ACER.

Sullivan, P. (2011). *Teaching mathematics: Using research-informed strategies*. Camberwell: ACER.

Sullivan, P., & Mornane, A. (2014). Exploring teachers' use of, and students' reactions to, challenging mathematics tasks. *Mathematics Education Research Journal*. doi:10.1007/s13394-013-0089-0.

Sullivan, P., Mousley, J., & Jorgensen, R. (2009). Tasks and pedagogies that facilitate mathematical problem solving. In B. Kaur (Ed.), *Mathematical problem solving* (pp. 17–42). Association of Mathematics Educators: Singapore/USA/UK World Scientific Publishing.

Sullivan, P., Youdale, R., & Jorgensen, R. (2010). A study of pedagogies for teaching mathematics in remote Australian indigenous community. In I. Synder & J. Nieuwenhuysen (Eds.), *Closing the gap? Improving outcomes in southern world societies* (pp. 204–216). Clayton: Monash University Press.

Swan, M. (2005). *Improving learning in mathematics: Challenges and strategies*. Sheffield: Department of Education and Skills Standards Unity.

Chapter 2
Large-Scale Test Data: Making the Invisible Visible

Gilah Leder and Sarah Lubienski

Introduction

> Information on pupil performance is key to the successful implementation of targeted education policies and it is not surprising that in the past two decades national tests have emerged as an important tool for providing a measure of educational achievement. (Figel 2009, p. 3)

Formal assessment of achievement has a long history. Official written examinations for selecting civil servants were already in use in China more than 3000 years ago (Kenney and Schloemer 2001). Educational assessment is a far more recent practice, with its genesis commonly traced to the nineteenth century. Over time, the development of large-scale, high-stakes testing and explorations of its results have proliferated. “Many nations,” wrote Postlethwaite and Kellaghan (2009), “have now established national assessment mechanisms with the aim of monitoring and evaluating the quality of their education systems across several time points” (p. 9). In some countries, the practice is limited to a number of core curriculum subjects but in others the testing regime is broad.

In this chapter, we confine our attention to large-scale tests used to measure mathematical progress and proficiency in two countries: Australia and the USA. We examine the aims, capacity, and limitations of the tests, what they can tell us about student performance and what can be learnt from an international comparison. In the Australian setting we focus primarily on the National Assessment Program—Literacy and Numeracy (NAPLAN). For the American context we have chosen to focus initially on the National Assessment of Educational Progress (NAEP). To supplement the findings from these tests limited reference is also made to three other tests: the (American) Early Childhood Longitudinal Study (ECLS), the Trends

G. Leder (✉)
Faculty of Education, Monash University, Clayton, Australia
e-mail: gilah.leder@monash.edu

S. Lubienski
Department of Curriculum and Instruction, University of Illinois, Urbana-Champaign, USA
e-mail: stl@illinois.edu

A. Bishop et al. (eds.), *Diversity in Mathematics Education*,
Mathematics Education Library, DOI 10.1007/978-3-319-05978-5_2,

in International Mathematics and Science Study (TIMSS), and the Programme for International Student Assessment (PISA).

How justified is our faith in using large-scale test results as a catalyst for change, for dealing constructively with diversity, and for enhancing social inclusion?

Early Developments

The USA

In a submission to America's National Assessment Governing Board (NAGB), Vinovskis (1998) sketched the motivations of early American advocates for the collection of comparative educational data as follows:

> Nineteenth-century reformers had an abiding faith that the compilation and display of numerical data not only would reveal the inherent regularities in behavior, but also would suggest possible options for making changes. They believed that if policymakers and the public were presented with the appropriate comparative data on social reforms such as education, they would soon want to improve their own policies accordingly. (Vinovskis 1998, p. 3)

These sentiments, we show in this chapter, still underpin—within America, Australia, and indeed more widely—contemporary preoccupations with large-scale testing and its putative benefits.

Much debate, political manoeuvring, and balancing of competing concerns and interests of local, state, and federal bodies preceded the creation and eventual introduction in the USA of the NAEP tests in the mid-1960s (see, e.g., Vinovskis 1998; Jones and Olkin 2004). The tests are now administered regularly to representative samples of students at grades 4, 8, and 12.

Australia

For many years, Australian states and territories ran their own numeracy and literacy testing programs. Although much overlap could be found in the assessment instruments used in the different states, there were also variations—some subtle, others substantial—in the tests administered. Finally, in 2008, a serious program of national testing was launched. The NAPLAN tests represented a significant turning point in Australia's educational system. For the first time, students in years 3, 5, 7, and 9, irrespective of their geographic location in Australia, sat for a common set of tests.

More about NAEP and NAPLAN

NAEP

Every 2–4 years, NAEP assesses the mathematics knowledge and attitudes of large, representative samples of US students at grades 4, 8, and 12 (roughly ages 9, 13, and 17, respectively). The Main NAEP mathematics results are reported via overall scale scores and proficiency levels (basic, proficient, advanced). Scale scores are also available for each of five mathematical strands: (1) number/operations; (2) measurement; (3) geometry/spatial sense; (4) data analysis, statistics, and probability; and (5) algebra/functions. Additionally, NAEP administers student and teacher surveys to collect information about students' mathematics attitudes and experiences, as well as teachers' backgrounds and classroom practices.

Several features complicate the analysis of NAEP data. The assessment uses multi-staged, stratified random sampling (geographic areas, then schools, and then students are selected). Since NAEP is designed to provide a snapshot of national achievement as opposed to providing feedback to individual students or schools, each student is administered only a subset of the NAEP items. This allows NAEP to monitor national performance on a rich variety of items without over-burdening individual students. These complications are addressed with several techniques, including the use of student sampling weights and imputed achievement values. For further information about the structure of NAEP data, see Johnson (1992) or Johnson and Rust (1992).

Reporting NAEP Data

The National Center of Education Statistics publishes general NAEP results for the USA and for key demographic subgroups. Scores are reported for the nation and each state—scores for individual students, teachers, and schools are not available. State tests—not NAEP—are typically used to track the progress of individual US students and schools.

NAEP scores for grades 4 and 8 are on a common (cross-grade) scale of 0–500. Since 2005, grade 12 scores are on a 0–300 scale. Most of the NAEP analyses discussed here were conducted via NAEP's web-based data tool available at http://nces.ed.gov/nationsreportcard/naepdata/. This tool allows users to examine mathematics achievement and survey data, and to make comparisons by demographic variables, including socioeconomic status (SES), gender, race/ethnicity, and home language. Raw NAEP data are also available to researchers via an application process.

NAPLAN

The NAPLAN numeracy tests contain multiple choice and supply response items. Their scope and content are informed by the Statements of Learning for Mathematics (Curriculum Corporation 2006) and cover four broad, and sometimes overlapping, numeracy strands: algebra, function, and pattern; measurement, chance, and data; number; and space. Since students have 40 or 50 minutes (depending on grade level) to complete a numeracy test paper, the content and scope of these high-stakes tests are inevitably limited.

Reporting the Data

Students' NAPLAN numeracy scores for years 3, 5, 7, and 9 are reported on a common scale, based on the Rasch model (see, e.g., Andrich 1988), so that any given scale score represents the same level of achievement over time. The use of a common scale for all domains measured enables changes in individual student achievement to be tracked, and provides a longitudinal dimension to the data. For each year level, the proportion of students with scores in the six proficiency bands deemed appropriate for that level is provided. For year 3, the bands are 1–6; for year 5, 3–8; for year 7, 4–9; and for year 9 bands 5–10.

Each year, summative results of the NAPLAN tests are published in some detail and are made available to the public. Particular attention is paid to the proportion of students meeting, or failing to meet, the specified minimum standards, and to mean NAPLAN scale scores. Individual student data are released to the relevant school attended, and the student's parents; unless special consent has been obtained, they are otherwise unavailable.

As well as the aggregated data, results are reported separately by: state/territory, gender, Indigenous status, language background status (language background other than English; LBOTE and non-LBOTE), geolocation (metropolitan, provincial, remote, and very remote), parental educational background, and parental occupation. These factors overlap with those frequently used as descriptors of equity/inequity. The categories are not mutually independent and can have a simple or compounding impact on students' NAPLAN scores. At the same time, as noted in Chapter 6, such data can, indirectly and subtly, reinforce prevalent stereotypes. More generally, as we show in this chapter, the publication of results in this way can highlight advantages and disadvantages linked to situational and external factors.

Beyond NAEP and NAPLAN

Australia and the USA also participate in two highly influential large-scale international tests of mathematics achievement: the TIMSS and the PISA. In both countries, students' results on the national and international tests attract much attention, within and beyond the educational community (see, e.g., Carnoy and Rothstein 2013; Thomson 2010; Thomson et al. 2011). In the most recent TIMSS tests (see Mullis, Martin, Foy and Arora 2012), the average scale score for Australian students at the fourth grade and eighth levels in TIMSS was somewhat below that of students in the USA—25 points at the former and 4 points at the latter grade level. On the PISA 2012 test, the mean score of Australian students (504) was above the international mean score; that of American students (481) was below the international average.

Because of differences in the scope of the tests and in the student groups to whom the large-scale tests are administered, complementary as well as overlapping data are yielded by the various testing instruments: NAEP, NAPLAN, TIMSS, and PISA.

Comparing the Design of NAPLAN, NAEP, TIMSS, and PISA

NAPLAN

As described above, the NAPLAN testing regime is aimed at students in years[1] 3, 5, 7, and 9. Participation in the tests is formally voluntary. Compliance, however, is high. For example, in 2011 approximately 95 % of the Australian year 3 cohort and 92 % of the year 9 cohort completed the NAPLAN numeracy tests. Thus, NAPLAN is best described as a census test. The NAPLAN numeracy tests provide information about student performance, both individually and for specific groups, on traditional mathematical tasks taught in class. There is no attempt to measure students' attitudes or to probe student knowledge not readily measured within the short time slot allocated to the test.

NAEP

NAEP is administered regularly to selected students in grades 4, 8, and 12, with the sample chosen to be nationally representative with respect to factors such as ethnicity, economic background, geographic location, and school size. Only a small percentage (less than 5 %) of US students in those grades participate. Still, the sample sizes are large, with roughly 175,000 at grades 4 and 8 and 50,000 at grade 12 participating in recent assessments.[2] Almost 100 % of selected fourth and eighth graders at public schools take the test, with participation rates lower for twelfth graders (83 %) and private school students (70–75 %).[3] The mathematical content assessed by the "Main NAEP" (discussed thus far and throughout the chapter) reflects current thinking about the most important curricular topics and includes a mix of multiple choice, short answer, and extended response items.[4]

TIMSS and PISA

The TIMSS tests, aimed at students in years 4 and 8, and the PISA tests administered to 15-year-old students, are restricted in Australia to "a light sample (of) about

[1] Terminology (years or grades) as used in the tests.

[2] Grade 4 and 8 samples are particularly large because they are selected to be representative of each US state and then aggregated to be nationally representative. The grade 12 sample, in contrast is simply nationally representative and therefore smaller.

[3] These and other details are available at: http://nationsreportcard.gov/math_2009/about_math.asp, http://nationsreportcard.gov/math_2011/about_math.asp, and http://nces.ed.gov/nationsreportcard/tdw/sample_design/.

[4] The "Long-Term Trend" NAEP (administered periodically with far smaller sample sizes) is a more traditional, multiple-choice test that tracks US students' mathematics and reading knowledge on the content considered important when it was begun in the early 1970s. We focus in this chapter on the more widely referenced and discussed "Main NAPLAN" data.

5 % of all Australian students at each year or age level" (Thomson 2010, p. 76). A similar sampling approach is used in the USA, with roughly 5000–8000 students (less than 1 %) in each cohort assessed (http://nces.ed.gov/timss/faq.asp; Carnoy and Rothstein 2013).

The aim of the TIMSS testing program is ambitious. Its scope is typically discussed in terms of three levels of the curriculum. These are:

> the intended curriculum (what society expects students to learn and how the system should be organised to facilitate this), the implemented curriculum (what is actually taught in classrooms, who teaches it and how it is taught) and the achieved curriculum (which is what the students have learned, and what they think about these subjects). (Thomson 2010, p. 76)

Thus, the TIMSS tests assess how well students can handle the work taught in class, as well as a range of broader situational and background factors.

The PISA tests have a different focus. The mathematical component of this test aims to assess not only how well students have mastered mathematical content but also how well they can apply that knowledge to real world settings.

National Tests: Intended Benefits

Australia

The reputed benefits of a national testing program were widely discussed and disseminated prior to the introduction of the NAPLAN tests. They continue to be repeated in official documents published by the Australian Curriculum Assessment Reporting and Curriculum Authority (ACARA). Advantages of the test agenda which continue to be lauded by its supporters mirror those commonly put forward in the wider literature: assessment consistency across different constituencies, increased accountability, and a general driver for improvement.

> The national tests, which replaced a raft of tests administered by Australian states and territories, improved the comparability of students' results across states and territories…
> Australians can expect education resources to be allocated in ways that ensure that all students achieve worthwhile learning during their time at school…
> All Australian schools benefit from the outcomes of national testing, with aggregated results made available through comprehensive reports at the national and school level, accessible on-line. (excerpts retrieved from ACARA 2011a)

USA

Similar to the goals of NAPLAN, NAEP is commonly referred to as "the nation's report card." Unlike the variety of state and district tests that continue to be administered in US schools, NAEP is the only ongoing assessment of students representing the US. Given NAEP's inclusion of demographic and instruction-related variables, NAEP is also designed to monitor US students' learning experiences and disparities in academic outcomes.

National Tests: Critics' Concerns

There are, inevitably, those who question the benefits of large and high-stakes testing programs and express unease about their impact. Their voices, too, must be acknowledged. Criticisms often expressed

> range from the reliability of the tests themselves to their impact on the well-being of children. This impact includes the effect on the nature and quality of the broader learning experiences of children which may result from changes in approaches to learning and teaching, as well as to the structure and nature of the curriculum. (Polesel et al. 2012, p. 4)

The tendency, in both Australia and the USA, to use published test data to make glib and indefensible comparisons between schools has also caused dismay. The push for teacher and administrator accountability based on standardized test results has attracted much condemnation and has led to recent, high-profile cases of "cheating" among the staff in some US and Australian schools (Bachelard 2011; Strauss 2013; Winerip 2013).

Additionally, some scholars argue that standardized tests are inappropriate measures of achievement that often hinder instead of help efforts toward equity. For example, Gutiérrez (2008) has argued that standardized tests are overly narrow measures and that repeatedly calling attention to achievement gaps between groups while ignoring within-group variation only serves to confirm stereotypes.

What *can* be Learned from the NAPLAN and NAEP Tests?

Our emphasis in this section is on productive ways of interpreting the data, of using the published results to raise questions and issues which warrant further investigation, and focusing on current inequities that cry out for positive interventions. We begin by listing a number of indicative examples taken from the NAEP and NAPLAN tests, and discuss what can, and cannot, be inferred from the results. Where useful, we refer to data beyond these tests.

In this way we not only examine the claims that large-scale tests contribute to increased accountability in the educational sectors and can serve as general drivers for improvement, but also highlight issues which merit reexamination and further explorations.

We start with NAEP and NAPLAN test data to examine the following questions:

- Has student mathematics achievement changed over time? Are any patterns observed unique to mathematics?
- Do we obtain any additional information if the test data are reported separately for different groups? In this chapter, we consider test outcomes by race/ethnicity (NAEP data only), language background, indigeneity (NAPLAN data only), gender, and geolocation. The impact of geolocation on students' achievement, in and beyond NAPLAN, is also addressed, and in some detail, in Chapter 3 of this volume.

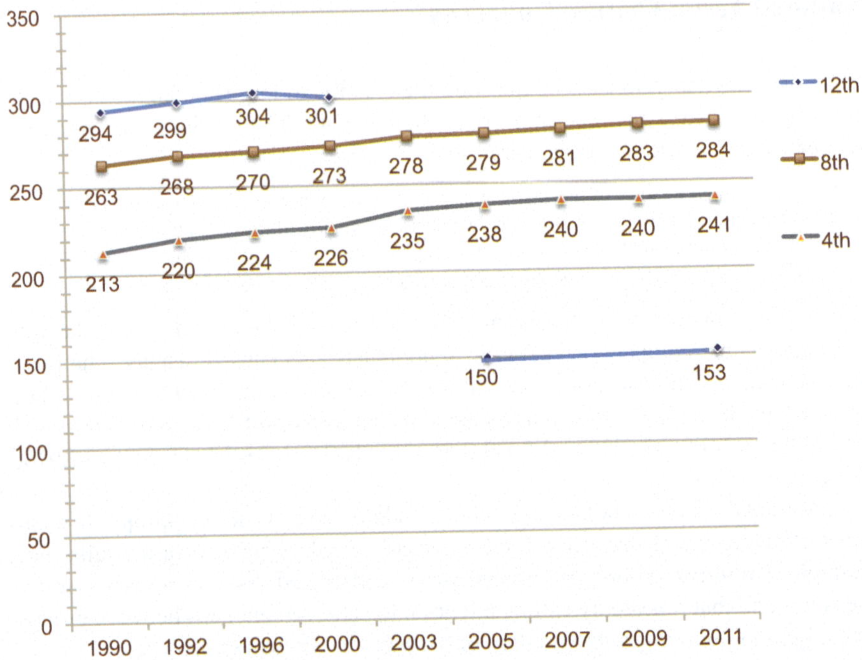

Fig. 2.1 Main NAEP mathematics results, 1990–2011, grades 4, 8, and 12

- What can we learn from item-level analyses?
- How can national survey data on student affect and instruction help us understand disparities between groups?

Achievement in Mathematics Over Time

Main NAEP Results 1990–2011

Since 1990, when the current Main NAEP Framework was established for grades 4 and 8, achievement has substantially increased at those grades. The standard deviation varies by year and grade level, but averages about 30 points at grade 4 and 36 points at grade 8. As shown in Fig. 2.1, increases are large, consisting of roughly 0.6–0.9 standard deviations. Gains at grade 12 are less evident, although the change in framework and reporting scale makes it difficult to draw conclusions about trends at that grade level. In contrast, during this same time period, reading scores increased only about 0.1–0.2 standard deviations at grades 4 and 8.

One of the most plausible explanations for the upward trend in mathematics is that the NCTM Standards movement began in 1989, and the framework for the 1990 NAEP assessment was designed to be aligned with those standards. That

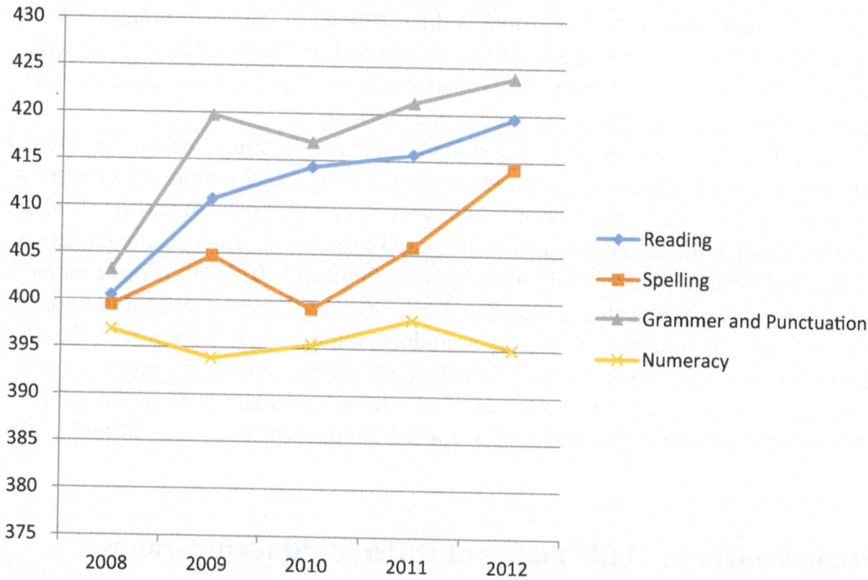

Fig. 2.2 NAPLAN Numeracy scores compared to three NAPLAN literacy measures

framework included topics that had typically not been included in the elementary and middle-school curriculum, such as probability, data analysis, and algebra. The NAEP framework also emphasized mathematical reasoning and communication more than prior assessment frameworks. The reforms influenced state and district standards and assessments (Usiskin 1993), as well as curriculum materials and teacher education programs. US students began having the opportunity to learn the content emphasized on NAEP, and this can be seen in the score gains during the 1990s and perhaps beyond. Another theory to explain gains made in the beginning of 2003 is that the new federal law, "No Child Left Behind," began to shift toward more high-stakes uses of NAEP and other standardized tests, which put pressure on educators to prepare students for such tests and to pay particular attention to the performance of traditionally underserved student groups.

Main NAPLAN Results for 2008–2012

For NAPLAN, the time span for monitoring test results is—at the time of writing this chapter—limited to 5 years.

At each of the year levels tested, only one numeracy test is administered, compared with four different tests for literacy: reading, writing, spelling, and grammar and punctuation. The mean test scores for four of the five national achievement scales at the year 3 level are shown in Fig. 2.2. The data for writing have been omitted because of a change to this test over the life of the NAPLAN testing regime, making it inappropriate to report longitudinal data from this test. Initially, for the period 2008–2010, the writing results were reported on the narrative writing scale; in 2011–2012 this was changed to the persuasive writing scale to reflect more effectively the range of what is required in the curriculum.

From Fig. 2.2, it can be seen that numeracy scores have remained steady, while literacy scores have generally increased over the 2008–2012 period. Space restrictions prevent detailed presentation of the year 5, 7, and 9 data. What can be said, however, is that the pattern evident at the year 3 level of relatively steady numeracy scores and somewhat increased scores between 2008 and 2012 for the three literacy measures is largely repeated at the other 3 year levels.[5] This is an intriguing finding. Are the different numeracy and literacy findings a function of the nature of the different tests and the way they capture curriculum content, or does the explanation lie elsewhere? Why, it should also be asked, do the achievement patterns for grade 4 students on the NAEP tests and year 3 students on the NAPLAN tests differ?

Achievement in Mathematics of Different Student Groups

NAEP Achievement by Race/Ethnicity and Sex

The impressive gains in NAEP scores since 1990 raise the question of whether all US students made similar gains, or if gaps between more and less advantaged students increased or decreased during this time period.

NAEP offers information on many different subgroups. The three largest racial/ethnic subgroups for which NAEP reports data are White, Black, and Hispanic students.[6] Figure 2.3 below reveals that White, Hispanic, and Black students all made fairly similar gains between 1990 and 2011. However, the steepest gains for Black and Hispanic students were made between 2000 and 2003, leading many politicians to credit the "No Child Left Behind" reforms for the narrowing of gaps. The patterns at grade 8 were similar.

There are additional patterns to note in Fig. 2.3. First, throughout the years, White students consistently scored 0.6−1.0 standard deviation higher than Hispanic and Black students. These are large, persistent disparities that merit continued attention. However, these same data show that such disparities are not inevitable. In fact, it took less than two decades for Black and Hispanic students to surpass a mean score of 220, which was the mean for White students in 1990. In other words, if White students had maintained their initial 1990 scores while mathematics learn-

[5] See, e.g., 2012 NAPLAN National Report, retrieved from www.nap.edu.au/verve/_resources/NAPLAN_2012_National_Report.pdf.

[6] Although there are differences of opinions about appropriate categories and terms to use to describe various groups in the USA, we use NAEP's school-reported categories and terms when describing NAEP results for racial and ethnic subgroups. Latino/a students are included as "Hispanic" and are generally not included in the "White" and "Black" categories. NAEP also includes a category for "2 or more races," but we do not report on that relatively small category here.

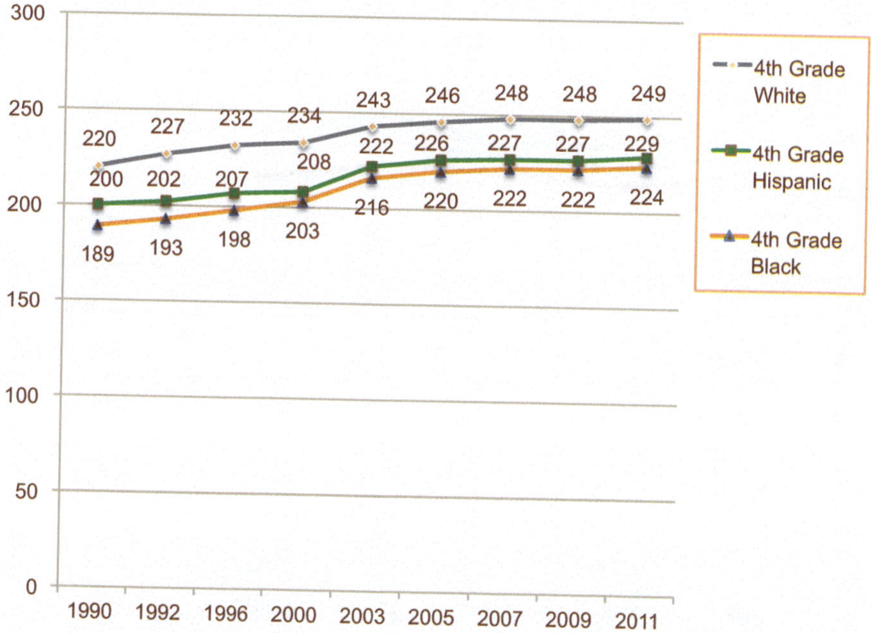

Fig. 2.3 Grade 4 NAEP mathematics results, 1990–2011, for Black, Hispanic, and White students

ing opportunities changed for Hispanic and Black students as they did during these decades, then achievement gaps would be closed. This provides important evidence about the power for instructional changes to raise students' achievement. However, the persistence of the disparities also cautions us that simply enhancing opportunities for everyone does not remove inequities or promote inclusivity.

Patterns for US girls and boys are very similar to those for race/ethnicity, although the size of the achievement gap is much smaller. Specifically, although NAEP achievement increased similarly for both girls and boys from 1990–2011, there was a consistent gap favoring males of roughly 0.1 standard deviations.

NAPLAN Achievement by Sex, Language, and Indigeneity

For the period 2008–2012, the numeracy NAPLAN scores for students in year 3 are shown overall, by sex, language background, and Indigeneity in Fig. 2.4.

A number of consistent patterns can be seen from the data in Fig. 2.4:

Language Background

For each of the years 2008–2012, LBOTE students consistently do better on the test (that is, have a higher mean score) than Non-LBOTE students. The stronger

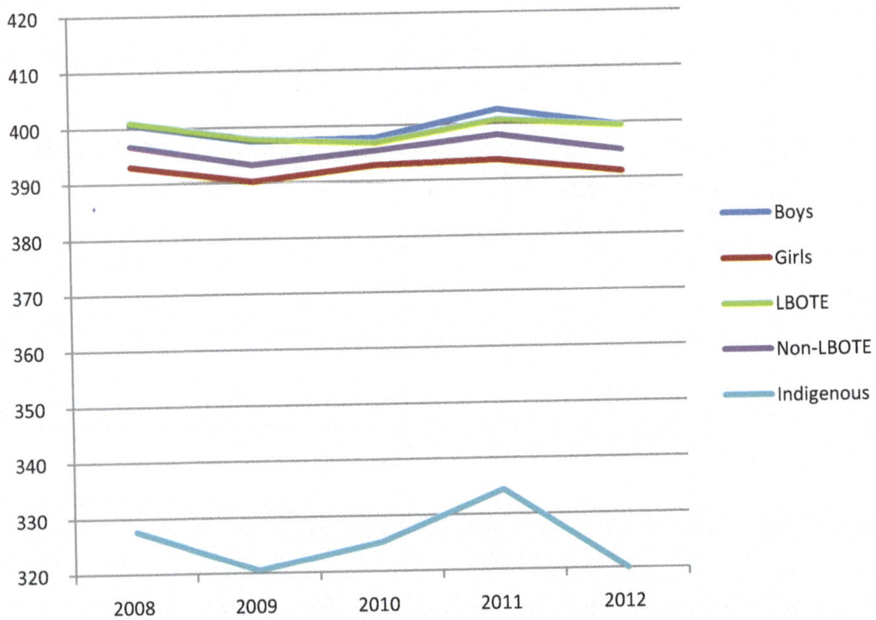

Fig. 2.4 NAPLAN scores by group. *LBOTE* language background other than English

performance of this group on the NAPLAN test is replicated on the years 5, 7, and 9 tests. Superficially, this result is counter-intuitive, given the common assumption in Australia that students whose native language is not English are disadvantaged educationally. On the other hand, a different picture emerges when the percentage of students at or above the national minimum standard is considered. Using that measure of achievement, in each year and at each grade level, a slightly higher proportion of non-LBOTE than LBOTE students do better. Data for the 2012 year 5 NAPLAN numeracy test are representative of the difference: 93.9 % of non-LBOTE students compared with 91.4 % of LBOTE students met or exceeded the national minimum standard score. Which of these two measures of performance is more helpful in determining which group needs special assistance?

Boys/Girls

For each of the years 2009–2012, boys as a group do better (that is, have a higher mean score) than girls as a group. This finding, too, is replicated for the NAPLAN tests at the other 3 year levels, and suggests that the issue of sex differences[7] in mathematics performance merits continued monitoring and examination.

[7] In this chapter, we use "sex differences" when it is clear that categorization is only based on biological factors. "Gender differences" are used when psychosocial or sociocultural factors may contribute to any difference found.

Table 2.1 2012 NAPLAN tests for years 3, 5, 7, and 9. Selected group data by sex

NAPLAN test	Mean		% at or below national min		% at or above national min		% in highest band	
	M	F	M	F	M	F	M	F
Year 3	399.5	391.2	15.2	15.8	93.3	93.5	14.4	10.3
Year 5	492.1	485.1	17.6	18.7	92.6	94.6	11	8
Year 7	543.7	532.4	19.7	22	93.5	94.1	12.4	7.9
Year 9	590	578.1	21.7	24.8	93.9	93.5	10.9	7

M = male, F = female

As already mentioned, NAPLAN data are also reported by proficiency band and percentage of students meeting, or failing to meet, a minimum specified standard. Such data, for 2012, are shown in Table 2.1.

Differences between the percentages of boys and girls who score at or above the prescribed national minimum score are very small.[8] In each year except year 9, a slightly higher percentage of girls reach this minimum level. But this group also includes students who are considered at risk with respect to mathematics achievement. "It should be noted that students who are performing at the National Minimum Standard may also require additional assistance to enable them to achieve their potential" (ACARA 2011b). Again, as shown in Table 2.1, the differences in the percentage of boys and girls whose score puts them below or at the minimum national standard are small, but at each year level the percentage of girls exceeds the percentage of boys in this category. Not to be ignored, either, is the difference in the proportions of boys and girls in the highest NAPLAN band, already evident at the year 3 level. Though not shown here, this difference is also found in the other years in which the NAPLAN tests have been administered. Can the observed differences be attributed to the design of the NAPLAN test, the nature and scope of its items, or should other factors and explanations beyond the test per se be invoked, for example, gender differences in preferred method of learning (see Chapter 5) or social attitudes about the learning of mathematics (Leder and Forgasz, Chapter 6)?

Indigeneity

From Table 2.2, it can further be seen that the mean score of Indigenous students on the NAPLAN test is consistently lower than that of non-Indigenous students—a finding once more replicated on the NAPLAN tests at years 5, 7, and 9. That reporting data at this crude level hides important inequities linked to the location of schools is illustrated unambiguously in Table 2.2.

[8] The differences reported are somewhat smaller than the differences found on this measure between the non-LBOTE and LBOTE students discussed in the previous section.

Table 2.2 Year 3 NAPLAN mean scores by indigeneity and geolocation

Year/group	2008	2009	2010	2011	2012
Metro non-indigenous	404	401.7	403	406	404.4
Indigenous	345.7	339.7	343.6	348.5	339.4
Provincial non-indigenous	392.3	387.4	388.4	390.3	385.6
Indigenous	339.2	334.3	336.5	341.8	330.7
Remote non-indigenous	377.5	375.3	380.8	378.1	371.9
Indigenous	305.7	287.4	307.4	313	290.8
Very remote non-indigenous	376.1	371.7	370.5	376	367
Indigenous	265.9	251.2	261.4	286.6	250.4

The impact of geolocation on the mean NAPLAN score is, it can be seen from Table 2.2, far greater for Indigenous than for non-Indigenous students. The numeracy score of non-Indigenous students at schools in very remote locations is still higher than that of Indigenous students at Metropolitan schools. The performance of both groups is affected by school locality, and dramatically so for Indigenous students in remote and very remote communities. These discrepancies in performance are also discussed in Chapter 3, this volume. They have troubled the Australian community, educators, and policy makers for many years. For credible explanations and interventions we must look beyond the NAPLAN test results, beyond one-off small studies, to longitudinal studies of Indigenous children. At this stage, such information is still limited.[9]

Performance Differences on Items

Analyses of student performance on individual test items can shed light on which types of mathematics items tend to have the largest disparities, identifying topics most in need of improved instruction for historically underserved students.

For example, on the 2003 grade 8 NAEP, although omit rates on multiple-choice items were less than 2 % for all racial/ethnic subgroups, the omit rate among Black or Hispanic students for items requiring an "extended-constructed response" was 20 %, or almost double the rate for White or Asian and Pacific Islander students (11 %) (Lubienski and Crockett 2007). Additionally, race- and ethnicity-related disparities were relatively small on basic computation items, but larger on nonroutine problems, such as those involving multiple steps or extraneous information.

For example, Fig. 2.5 displays an item involving an extraneous number on which there were particularly large disparities between the performance of Black and White students. Although almost two-thirds of White students correctly chose

[9] The first large-scale longitudinal survey of (Australian) Indigenous students [LSIC], also known as *Footprints in time*, began in 2008. The survey is conducted under the auspices of the Department of Families, Housing, Community Services and Indigenous Affairs. (Retrieved from http://www.fahcsia.gov.au/about-fahcsia/publications-articles/research-publications/longitudinal-data-initiatives/footprints-in-time-the-longitudinal-study-of-indigenous-children-lsic).

Pat has 3 fish bowls.
There are 4 plants and 5 fish in each bowl.
Which gives the total number of fish?

A) 3 + 5
B) 3 x 4
C) 3 x 5
D) 3 + 4 + 5

Fig. 2.5 Grade 4 NAEP item involving extraneous information. (Performance data on this and other released items are available via the NAEP Question Tool (http://nces.ed.gov/nationsreportcard/itmrlsx/))

James makes 12 pizzas.
He puts 4 pizzas on trays like this.

Which of these shows how James could work out the number of trays he needs?

A) 12 ÷ 4
B) 12 x 4
C) 12 – 4
D) 12 + 4

Fig. 2.6 NAPLAN item

answer C, only 37 % of Black students did so. Black students (43 %) were almost twice as likely as White students (22 %) to choose option D, which involves combining the three given numbers in the problem.

Overall, analyses of these and other NAEP items suggest that Black and other traditionally under-served students tend to have relatively few opportunities to solve nonroutine mathematics problems that require reasoning and problem-solving skills (Lubienski and Crockett 2007). This is particularly troubling given the two decades of reform in the USA emphasizing student reasoning and dedicated to promoting "mathematical power for all students" (NCTM 1989).

Inspection of NAPLAN data is also instructive. The choice of an item common to NAPLAN papers administered at 2-year levels allows comparisons to be made within and across students' performance at 2-year levels. Performance data for one such item common to the year 3 and year 5 2010 NAPLAN paper and shown in Fig. 2.6, are presented in Table 2.3.

Table 2.3 Results by group (% of students with the correct answer) for an item common to the NAPLAN papers (in 2010) for years 3 and 5

Question/group	Year 3 Q 24 (%)	Year 5 Q 18 (%)	Difference (%)
All	44.1	75.5	31.4
Boys	46.1	76.8	30.7
Girls	42.0	74.2	32.2
LBOTE	48.7	78.9	30.2
Indigenous	26.9	51.6	24.7

LBOTE language background other than English

From Table 2.3, it can be seen that in each group, a higher percentage of year 5 than year 3 students answered the question correctly. At both year levels, the LBOTE group had the highest percentage of correct answers; the Indigenous group the lowest. The increase in the percentage of students with the correct answer was less for the Indigenous group (~25 %) than for the other groups (≥30 %). By year 5, about half the group of Indigenous students were able to answer this question correctly, compared with about three-quarters of the students subsumed under the other categories used.

For the year 3 paper (see Fig. 2.6), B (multiply instead of divide) was the most common distracter chosen by three of the groups: Boys (22.2 %), Girls (27.1 %), and LBOTE students (23.3 %). D (add instead of divide) was the most common distracter chosen by the Indigenous group (26.1 %). On the year 5 paper, B was the most common distracter chosen by all groups.

Beyond Achievement—National Data on Affect and Instruction

NAEP goes beyond achievement measures and includes survey data from students and teachers. The following are just a few of the examples of what those data tell us about race/ethnicity-, SES-, and gender-related disparities in US mathematics education.

The teacher survey data reveal patterns in teacher qualifications. For example, the 2011 NAEP data indicate that Black US students are more than twice as likely as White students to have a teacher who entered teaching through an alternative certification route (as opposed to graduating from a standard teacher education program). Specifically, as shown in Fig. 2.7, while only 6 % of White fourth graders had a teacher who entered the profession through an alternate route, 13 % of Black fourth graders, and 16 % of Hispanic fourth graders had such a teacher. At grade 8, these percentages were 13 % for White students, 27 % for Black students, and 24 % for Hispanic students. Given concerns about the abbreviated nature of some alternate certification programs and the high dropout rates of teachers who are certified through such programs, these data point toward a potential source of inequity in the US education system (Heilig and Jez 2010; Laczko-Kerr and Berliner 2002).

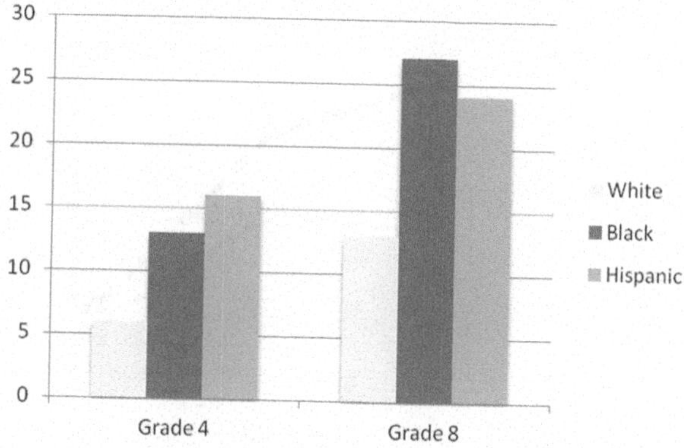

Fig. 2.7 Percentage of White, Black, and Hispanic students with a teacher who entered teaching through an alternative certification program, grades 4 and 8

NAEP student survey data also shed light on students' attitudes toward mathematics. For example, among 2011 fourth-graders in the NAEP sample, boys (42 %) were more likely than girls (36 %) to report that mathematics is "always or almost always" their favorite subject. Additionally, on average, Black students (44 %) were more likely than White students (35 %) to report that mathematics is their favorite subject.

Delving Further—Gender and Mathematics Confidence, Interest, and Achievement

Although the cross-sectional NAEP data can reveal patterns such as those noted above, longitudinal data are needed to shed light on the development and impact of such patterns over time. Hence, the USA periodically commissions longitudinal data collection efforts. The ECLS-Kindergarten Class of 1998–1999 (ECLS-K) is one such effort. ECLS-K followed a sample of US students from kindergarten through grade 8, with student, teacher, and parent data collected in grades K, 1, 3, 5, and 8. The ECLS-K sample began with approximately 21,000 kindergarten students from 1277 schools.

As just one example of ECLS-K analyses, Fig. 2.8 presents the standardized gender differences in mathematics interest, confidence, and achievement as the same group of students progresses through third, fifth, and eighth grades. The *dotted lines* in the figure raise a caution about changes in the confidence and interest variables that occurred at grade 8, making it difficult to draw conclusions about how those two factors really changed between grades 5 and 8 (Lubienski et al. 2012). However, the ECLS-K data indicate that the third and fifth-grade gender gaps in confidence are larger than the actual gap in achievement, suggested that girls are more insecure (or boys are perhaps overly secure) in their mathematics knowledge than is warranted. The data also show that the gaps in interest are actually smaller than the gaps in either confidence or achievement (Lubienski et al. 2013).

The longitudinal nature of the ECLS-K data allows us to look at relationships among these variables over time. Although these analyses are ongoing (Ganley and Lubienski, in preparation) results suggest that third-grade mathematics achievement

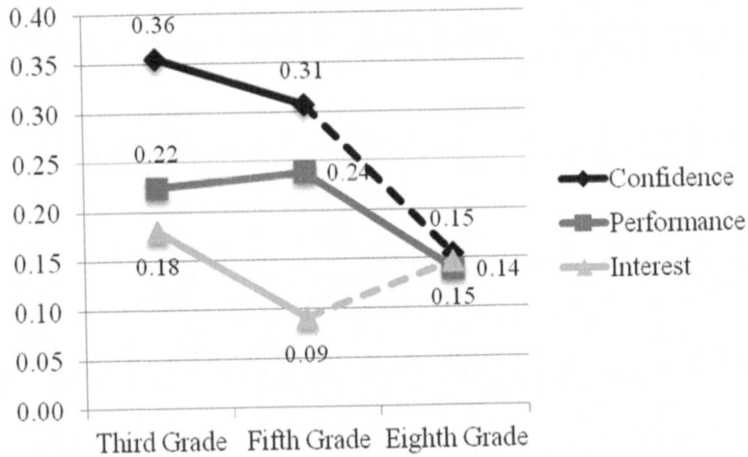

Fig. 2.8 Effect sizes for gender differences (male–female) in interest, confidence, and achievement

predicts later gains in achievement, confidence, and interest. However, third-grade interest appears to be a weak predictor, at best, of gains in achievement and confidence from third to eighth grades. The ECLS-K data point to the importance of enhancing young girls' mathematics achievement and confidence.

The Limits of Large-Scale Assessments

As noted above, some critics see little value in standardized tests. Certainly, such tests are but one tool for assessing student knowledge and skills, and we must remember that they are "pencil-and-paper, point-in-time, timed tests. The mathematical content covered…includes only that what can be assessed in this way, representing only a slice of the curriculum" (ACARA 2011b, p. 6).

But even those who see value in standardized assessments and analyze the resulting data run into specific constraints that must be acknowledged. Here, we discuss three such constraints.

First, there are often frustrating limits to the variables available—even for assessments, such as NAEP, which include student and teacher questionnaires. NAEP's self-reported survey items rarely get at the nuances of what is important in a mathematics classroom (e.g., what is actually taught and how).

Second, political and academic factors can sway decisions about the inclusion of specific mathematical content or survey items at particular grade levels or years. In longitudinal studies, this can make it difficult for researchers to track trends as students progress through school (as in the ECLS-K example above, in which the mathematical confidence and interest survey questions changed from grade 5 to grade 8). In cross-sectional surveys, changes in survey questions make it impossible to track national trends over time. As one example, between 1996 and 2003, NAEP asked US students about their level of agreement with the statements, "Learning math is mostly memorizing facts" and "There is only one correct way to solve a math problem." Students' level of agreement with these statements consistently differed by demographics, with high-SES, White, and Asian American students less likely to agree with the statements than others, indicating that these students had more opportunities to engage in genuine mathematical activity (Lubienski and Crockett 2007). Moreover, agreement with those statements was a persistent, negative correlate of achievement in multilevel regression models, even after controlling for student- and school-level demographics (Lubienski et al. 2008). Hence, tracking these variables over time provided an important window into students' mathematics learning opportunities and affective outcomes. Unfortunately, given political pressures to shorten and/or refocus the student surveys, the variables disappeared from NAEP in 2003.

Finally, analyses of national assessment data often raise but do not answer questions about *why* patterns exist as they do. Student, teacher, and parent surveys can point to persistent correlates of achievement, but causal conclusions are rarely warranted, particularly if only standard analysis techniques are used. However, this and other limitations of national assessments need not halt researchers' attempts to use these data to improve mathematics education and to further equity.

Increasing the Usability of Datasets in Mathematics Education Research

Given the massive resources invested in national and state assessments, it makes sense for the mathematics education community to consider ways in which the limitations of existing datasets might be overcome and to maximize our use of the data, while also being mindful that standardized tests are but one measure of student learning. In this final section, we take stock of what practitioners and researchers can learn from large-scale assessments, as well as steps that can help minimize the drawbacks and maximize the benefits of these data.

What can Practitioners, Curriculum Developers, and Educational Systems Apply?

Set Clear Instructional Targets

Item-level analyses can pinpoint the mathematics that students do and do not know, including which problems most students can and cannot solve, and which problems have the largest disparities between groups. This information can inform both textbook writers and teachers, as they strive to address curricular areas in need of additional attention. Hence, it is important for item-level analyses to be systematically conducted and reported (e.g., see Kloosterman and Lester 2007 for one such example).

Advance Mathematics Education Reform

National assessments can be highly effective levers of reform, particularly when assessments are closely aligned with the reformers' vision for curriculum and instruction and if the results have consequences for students and/or schools (Lubienski 2011). If a high-stakes assessment is narrow and highly specified, it can quickly cause a marked narrowing of the curriculum. However, if a mathematics assessment emphasizes nonroutine problem solving, reasoning, and writing mathematical solutions, then "teaching to the test" can actually increase students' opportunities to engage with genuine mathematics. Moreover, such assessments can promote equity by prompting all schools—not only those of traditionally advantaged students—to promote mathematical reasoning and problem solving for all students in all classes.

Promote Equity Through Accountability

Although "accountability" in conjunction with high-stakes testing is often viewed negatively, there are some ways in which such accountability may benefit students. In the American context, although there are many unpopular aspects to the US "No Child Left Behind Act," one potential benefit is that schools are now required to analyze and address their achievement gaps related to students' family income, race/ethnicity, and gender. In Australia, NAPLAN test data serve a similar function: "to inform future policy development, resource allocation, curriculum planning and, where necessary, intervention programs" (ACARA 2011a). The degree to which some schools have responded by narrowing the curriculum and teaching to the test is a concern, but one that may be overcome with assessments that are broad enough to deter instead of reward this narrowing.

What can Researchers Apply?

Monitor Education Policy and Reform

National assessments can help monitor reforms such as the standards promoted by the National Council of Teachers of Mathematics in the USA (1989, 1991, 2000) or the development of the Australian Curriculum underpinned by the Melbourne Declaration on Educational Goals for Young Australians (the Ministerial Council on Education, Employment, Training and youth Affairs, December 2008).[10] Specifically, large-scale assessments can allow researchers to track student achievement over time, document instructional shifts over time, and shed light on the impact of

[10] see http://www.mceecdya.edu.au/verve/_resources/national_declaration_on_the_educational_goals_for_young_australians.pdf.

reforms on student outcomes. However, in order to do so, assessments must include relevant measures of instruction and student outcomes.

When studies of education policy are conducted by those outside of mathematics education, the ways in which problems are framed and data are interpreted are often inconsistent with current scholarship in our field. Hence, it is important for mathematics education researchers, in particular, to study the implementation and effects of new policies and to communicate with policy makers (Lubienski 2008). The American Statistical Association (2007) noted that there are more scholars from fields such as sociology and economics than mathematics education who pursue analyses of mathematics-related datasets such as NAEP and TIMSS (American Statistical Association 2007). Scholars with expertise in mathematics education and equity are especially needed to interpret findings in ways that will promote instead of hinder efforts to improve mathematics instruction and equity, and to inform the translation of mathematics data into promising policies.

Understand Relationships Between Student Opportunities and Outcomes

We already know that gaps based on race, SES, and gender exist, and studies that simply confirm those gaps are of questionable value. However, further analyses can illuminate when gaps begin, under what conditions they widen or narrow, and what consequences underserved students ultimately face because of the gaps (e.g., see Reardon and Robinson 2007). Concerns about causality are becoming easier to address with the development of quasi-experimental methods, including propensity score matching and regression discontinuity.[11] Additionally, some researchers are now combining multiple datasets to shed new light on student opportunities and outcomes, such as Hogrebe and Tate's (2012) work with geo-spatial data, examining ways in which geography influences educational opportunities.

Inform Interventions

Detailed analyses of gaps can help mathematics education scholars more effectively target their efforts to promote equity, illuminating which groups to target for particular interventions. For example, it is important to know that women in both Australia and the USA are currently more likely than men to graduate from university

[11] Propensity score matching allows researchers to match "treatment" and comparison groups on a large set of covariates in order to limit the potentially confounding effects of those covariates. Regression discontinuity designs are useful when a cut-score divides students (or others) into treatment and comparison groups, such as when schools use income, language ability, or a test score to determine which students should receive particular services or opportunities. See Stuart (2007) for further discussion of both of these methods. See Robinson (2010) for an application of regression discontinuity to the issue of mathematics test translations for English learners in the USA.

or college but less likely to choose mathematics-related fields. In contrast, African American and Latina/Latino students are, in general, less likely to get to college than White students but roughly as likely to major in mathematics-related fields once there (National Center for Education Statistics (NCES) 2006). Information about which groups and mathematical topics are most in need of targeting at particular grade levels is important for mathematics educators to draw from as they design equity-focused interventions which promote inclusivity.

Counter Deficit Perspectives

Although data can be used to confirm and promote harmful stereotypes, they can also be used to counter the idea that gaps are inevitable and fixed. For example, scholars can balance discussions of gaps between groups with noting the substantial overlap in the achievement distributions of those groups as well as the complexities of group membership (e.g., students who identify as more than one race or ethnicity). Scholars can also emphasize the temporal nature of achievement disparities. In both Australia and the USA, there are many, and often well publicized, examples to counter notions that girls or racial and ethnic minority groups simply are not good at mathematics.

Future Directions

Underlying many of the points made above is the fact that the usefulness of large-scale assessment data would be enhanced if a richer set of contextual factors and outcomes were utilized. We close with some questions for the mathematics education community to consider:

- Can mathematics education researchers come together and identify survey items that would enhance the usefulness of large-scale assessments?
- Can we develop a pool of items that have proven useful in smaller-scale studies and that would be promising to implement at national and international levels?
- Could we build items from mathematics education research on factors that promote student learning and equity (e.g., particular types of instruction or specific aspects of teacher knowledge)?
- Can we better capture outcomes that matter to those concerned about mathematics education and equity, such as students' ability and inclination to use mathematics to address injustices?

As should be clear from the questions above, smaller-scale studies can and should inform our efforts to improve the usefulness of large-scale data. The relationship between large-scale and smaller-scale studies should be an iterative process that leads to insights about ways to increase equity in students' mathematical experiences and outcomes.

References

American Statistical Association. (2007). *Using statistics effectively in mathematics education research*. Alexandria, VA: Author. http://www.amstat.org/education/pdfs/UsingStatisticsEffectivelyinMathEdResearch.pdf. Retrieved from 19 June 2014.

Andrich, D. (1988). *Rasch models for measurement*. Beverly Hills: Sage.

Australian Curriculum Assessment Reporting and Curriculum Authority (ACARA). (2011a). Why NAP. ACARA. http://www.nap.edu.au/about/why-nap.html. Retrieved from 6 March 2012.

Australian Curriculum Assessment Reporting and Curriculum Authority (ACARA). (2011b). National Assessment Program: Standards. http://www.nap.edu.au/results-and-reports/how-to-interpret/standards.html. Retrieved from 7 Feb 2012.

Bachelard, M. (2011). Is this school incredibly good, or just incredible? *The Age*, March 6, 2011. (http://www.theage.com.au/national/education/is-this-school-incredibly-good-or-just-incredible-20110305-1bix3.html).

Carnoy, M., & Rothstein, R. (2013). *What do international tests really show about U.S. student performance?* Washington, DC: Economic Policy Institute.

Curriculum Corporation. (2006). Statements of learning for mathematics. http://www.mceetya.edu.au/verve/_resources/SOL_Maths_Copyright_update2008.pdf. Retrieved from 16 Dec 2010.

Figel, J. (2009). Preface. In *National testing of pupils in Europe: Objectives, organisation and use of results*. Brussels: Education, Audiovisual and Culture Executive Agency—Eurydice. http://www.eurydice.org. Retrieved from 27 March 2012.

Gutiérrez, R. (2008). A "gap-gazing" fetish in mathematics education? Problematizing research on the achievement gap. *Journal for Research in Mathematics Education, 39*(4), 357–364.

Heilig, J. V., & Jez, S. J. (2010). *Teach for America: A review of the evidence*. Boulder and Tempe: Education and the Public Interest Center & Education Policy Research Unit. (http://epicpolicy.org/publication/teach-for-america).

Hogrebe, M., & Tate, W. F. (2012). Place, poverty, and algebra: A statewide comparative spatial analysis of variable relationship. *Journal of Mathematics Education at Teachers College, 3*, 12–24.

Johnson, E. G. (1992). The design of the National Assessment of Educational Progress. *Journal of Educational Measurement, 29*(2), 95–110.

Johnson, E. G., & Rust, K. F. (1992). Population inferences and variance estimation for NAEP data. *Journal of Educational Statistics, 17*(2), 175–190.

Jones, L. V., & Olkin, I. (2004). *The Nation's report card: Evolution and perspectives*. Bloomington: Phi Delta Kappa Educational Foundation.

Kenney, P. A., & Schloemer, C. G. (2001). Assessment of student achievement, overview. In S. Grinstein & S. I. Lipsey (Eds.), *Encyclopedia of mathematics education* (pp. 50–56). New York: RoutledgeFalmer.

Kloosterman, P., & Lester, F. (Eds.) (2007). *Results and interpretation of the 2003 math assessment of NAEP*. Reston: National Council of Teachers of Mathematics.

Laczko-Kerr, I., & Berliner, D. C. (6 Sept 2002). The effectiveness of "Teach for America" and other under-certified teachers on student academic achievement: A case of harmful public policy. *Education Policy Analysis Archives, 10*(37), 34–38. (http://epaa.asu.edu/ojs/article/view/316).

Lubienski, S. T. (2008). On "gap gazing" in mathematics education: The need for gaps analyses. *Journal for Research in Mathematics Education, 39*(4), 350–356.

Lubienski, S. T. (2011). Mathematics education and reform in Ireland: An outsider's analysis of Project Maths. *Bulletin of the Irish Mathematical Society, 67*, 27–55.

Lubienski, S. T., & Crockett, M. (2007). NAEP mathematics achievement and race/ethnicity. In P. Kloosterman & F. Lester (Eds.), *Results and interpretation of the 2003 math assessment of NAEP* (pp. 227–260). Reston: National Council of Teachers of Mathematics.

Lubienski, S. T., & Gutiérrez, R. (2008). Bridging the "gaps" in perspectives on equity in mathematics education. *Journal for Research in Mathematics Education, 39*(4), 365–371.

Lubienski, S. T., Lubienski, C., & Crane, C. C. (2008). Achievement differences among public and private schools: The role of school climate, teacher certification and instruction. *American Journal of Education, 151*(1), 97–138.
Lubienski, S. T., & Ganley, C. M., & Crane, C. C. (2012). *Unwarranted uncertainty: Gender patterns in early mathematical confidence, interest and achievement.* Paper presented at the American Educational Research Association, Vancouver.
Lubienski, S. T., Robinson, J. P., Crane, C. C., & Ganley, C. M. (2013). Girls' and boys' mathematics achievement, affect and experiences: Findings from ECLS-K. *Journal for Research in Mathematics Education, 44*(4), 634–645.
Ministerial Council on Education, Employment, Training and youth Affairs. (Dec 2008). Melbourne declaration on educational goals for young Australians. http://www.mceecdya.edu.au/verve/_resources/national_declaration_on_the_educational_goals_for_young_australians.pdf. Retrieved from 16 Dec 2010.
Mullis, I. V. S., Martin, M. O., Foy, P., & Arora, A. (2012). *TIMSS 2011 international results in mathematics.* USA: TIMSS & PIRLS International Study Center and The Netherlands: IEA.
National Center for Education Statistics. (2006). *Digest of education statistics: 2005* (NCES 2006-030). http://nces.ed.gov/programs/digest/d05/. Retrieved from 23 May 2007.
National Council of Teachers of Mathematics. (1989). *Curriculum and evaluation standards for school mathematics.* Reston: National Council of Teachers of Mathematics.
National Council of Teachers of Mathematics. (1991). *Professional standards for teaching mathematics.* Reston: National Council of Teachers of Mathematics.
National Council of Teachers of Mathematics. (2000). *Principles and standards for school mathematics.* Reston: National Council of Teachers of Mathematics. (http://standards.nctm.org/).
Polesel, J., Dulfer, N., & Turnbull, M. (2012). *The experience of education: The impacts of high stake testing on students and their families.* Rydalmere: University of Western Sydney, The Whitlam Institute. (http://www.whitlam.org/__data/assets/pdf_file/0008/276191/High_Stakes_Testing_Literature_Review.pdf).
Postlethwaite, T. N., & Kellaghan, T. (2009). *National assessment of educational achievement.* Paris: UNESCO, International Institute for Educational Planning. (http://www.iiep.unesco.org/fileadmin/user_upload/Info_Services_Publications/pdf/2009/EdPol9.pdf).
Reardon, S. F., & Robinson, J. P. (2007). Patterns and trends in racial/ethnic and socioeconomic academic achievement gaps. In H. A. Ladd & E. B. Fiske (Eds.), *Handbook of research in education finance and policy* (pp. 499–518). Mahwah: Erlbaum.
Robinson, J. P. (2010). The effects of test translation on young English learners' mathematics performance. *Educational Researcher, 39*(8), 582–590.
Strauss, V. (2013). Atlanta test cheating: Tip of the iceberg? *Washington Post* (http://www.washingtonpost.com/blogs/answer-sheet/wp/2013/04/01/atlanta-test-cheating-tip-of-the-iceberg/).
Stuart, E. A. (2007). Estimating causal effects using school-level data sets. *Educational Researcher, 36*(4), 187–198.
Thomson, S. (2010). *Mathematics learning: What TIMSS and PISA can tell us about what counts for all Australian students.* Paper presented at the ACER Research conference: Teaching mathematics? Make it count. (http://research.acer.edu.au/research_conference/RC2010/17august/6/).
Thomson, S., De Bortoli, L., Nicholas, M., Hillman, K., & Buckley, S. (2011). *Challenges for Australian education.* Results from PISA 2009. (http://www.acer.edu.au/documents/PISA-2009-Report.pdf).
Usiskin, Z. (1993). What changes should be made for the second edition of the NCTM Standards? *University of Chicago School Mathematics Project Newsletter. Winter,* 6–11.
Vinovskis, M. A. (1998). *Overseeing the Nation's report card.* Paper prepared for National Assessment Governing board. (http://www.nagb.org/content/nagb/assets/documents/publications/95222.pdf).
Winerip, M. (30 March 2013). Ex-schools chief in Atlanta is indicted in testing scandal. *The New York Times.* (http://www.nytimes.com/2013/03/30/us/former-school-chief-in-atlanta-indicted-in-cheating-scandal.html).

Chapter 3
Impact of Geographical Location on Student Achievement: Unpacking the Complexity of Diversity

Debra Panizzon

Introduction

Globally, there are many political imperatives to address the diverse needs of students in order to overcome the inequity that is clearly present in many societies. Examples of policy documents include the *United Nations Declaration on the Rights of the Child* (United Nations Convention on the Rights of the Child (UNCRC 1990), *No Child left Behind* (US Department of Education 2001), and in Australia, the *Melbourne Declaration on Educational Goals for Young Australians* (Ministerial Council on Education, Employment, Training and Youth Affairs (MCEETYA 2008)). In brief, these documents articulate that *all* students regardless of socioeconomic and cultural backgrounds, learning abilities, or geographical location should be provided with equal opportunities to access learning and so achieve their academic potential. While much of this may be achieved through schooling, the degree to which this is achievable is dependent on a complex interplay between the school, parents, and the wider community.

The reality though is that most of the responsibility in catering for this range of diversity in education is targeted at schools and individual classrooms. As a result, it is classroom teachers who become responsible for meeting the needs of these students by employing a variety of inclusive practices and pedagogies. While there is considerable research around the area of inclusivity, in some countries much of this appears to be focused on teaching students with special needs (i.e., learning disabilities). Challenging this view, Ainscow (2007) highlights the international reform around inclusivity as a means of supporting and welcoming "diversity amongst all learners" (p. 3).

In this chapter, student diversity is considered at a broader scale around geographical location using Australia as a case study. Initially, the Australian context is described, followed by an analysis of data from the Program for International Student Assessment (PISA) and National Assessment Program Literacy and Numeracy (NAPLAN) tests in order to explore diversity in relation to student achievement. With patterns and trends identified, the research literature is reviewed to extract

D. Panizzon (✉)
Faculty of Education, Monash University, Clayton, Australia
e-mail: debra.panizzon@monash.edu

A. Bishop et al. (eds.), *Diversity in Mathematics Education*,
Mathematics Education Library, DOI 10.1007/978-3-319-05978-5_3,

the possible factors impacting variation in student achievement across geographical location. Finally, consideration is given to ways of enhancing inclusivity in mathematics while bearing in mind the complexity of challenges faced by teachers in their particular teaching contexts.

The Australian Context

Australia is a western industrialized country with a current estimated population of 22 million (Australian Bureau of Statistics (ABS 2009)). It is also one of the most urbanized countries in the world (State of the Environment Advisory Council 1996) with cities and larger centres located around the coastline leaving large areas of the continent dotted with towns of varying population sizes. Life in many of these smaller towns is vibrant, dynamic, and in some cases, increasingly cosmopolitan as city dwellers seek a "sea change" or "tree change" life style. However, distances between these small towns are often substantial with the number of services (e.g., medical practitioners, hospitals, ambulance contact, financial institution) declining with increasing distance from cities or regional centres. So, while living in these small rural towns might at first appear idyllic, the research literature for the Australian context identifies major difficulties for inhabitants in many communities in accessing essential services and higher levels of education (Squires 2003).

In general, students in most small rural towns enrol in their local schools for their primary education (i.e., 5–11 years of age) but may travel substantial distances to attend secondary schools located in larger regional centres. While most students might travel 2–3 h to and from school on a daily basis, others become boarders at schools in cities returning home for their holidays.

Meeting the needs of these rural students is a concern for the Australian government as articulated in the Melbourne Declaration on Educational Goals for Young Australians (Ministerial Council on Education, Employment, Training and Youth Affairs (MCEETYA 2008), p. 15):

> For Australian schooling to promote equity and excellence, governments and all school sectors must improve educational outcomes for Indigenous youth and disadvantaged young Australians and encourage them, their families and their communities to hold high expectations for their education.

In particular, the Melbourne declaration refers to the government's commitment to "close the gap" for young Indigenous Australians, support disadvantaged students (including those living outside of major cities), and focus on school improvement in low socioeconomic communities. While this is positive, the degree of success in meeting these goals is questionable, with evidence currently available suggesting there has been minimal improvement if at all.

Before exploring some of this evidence, it is important to highlight a confounding issue in Australia when interpreting the available information. As demonstrated above, there is a range of terminologies for describing population centres in Australia including *rural*, *small centre*, *regional centre,* or *remote*. In fact, so diverse are

Table 3.1 Categories of the MCEETYA schools geographic location classification (MSGLC)

Major zones	Category	Criteria	
Metropolitan zone	1.1	State capital city regions	State capitals (except Hobart, Darwin)
	1.2	Major urban statistical districts	Pop. > 100,000
Provincial zone	2.1.1	Provincial city statistical districts	Pop. > 50,000
	2.1.2	Provincial city statistical districts	Pop. 25,000–49,999
	2.2.1	Inner provincial areas	CD ARIA[a] index value ≤ 2.4
	2.2.2	Outer provincial areas	CD ARIA index value > 2.4 and ≤ 5.92
Remote zone	3.1	Remote areas	CD ARIA index value > 5.92 and ≤ 10.53
	3.2	Very remote areas	CD ARIA index value > 10.53

Pop. Population

[a] Accessibility and Remoteness Index of Australia (ARIA) is an accessibility and remoteness value between 0 and 15 with increasing value demonstrating access to fewer basic services. (Data from Jones 2004)

these terminologies that MCEETYA developed the Schools Geographical Location Classifications (SGLC) (Jones 2004) as a classification standard. Essentially, the MCEETYA SGLC (MSGLC) consists of eight categories (see Table 3.1).

The first four categories are based solely upon population while the last four categories reflect the ARIA for any centre using criteria, such as the physical road distance to the nearest service centre and the proximity to basic services (e.g., hospital, doctor, dentist) (Jones 2004). However, given the complexity of this table the categories are often collapsed within the literature into four groupings to include *Metropolitan Zone* (1.1, 1.2), *Provincial* City (2.1.1, 2.1.2), *Provincial Area* (2.2.1, 2.2.2), and *Remote Zone* (3.1, 3.2).

The importance of this classification is that it provides a degree of consistency and a common language for identifying population centres in Australia. The problem is that it is not helpful when analyzing the literature prior to the introduction of the MSGLC or when considering international literature. Without a consistent classification it makes wide-scale comparisons difficult and care is required around the interpretation of findings.

Impact of Geographical Location on Student Achievement

Given the Australian context described above, this section considers: *Does diversity around student achievement exist? If so, how extensive is this issue in Australia in mathematics and numeracy?* Access to large-scale international data sets, such as those provided by PISA, highlight that in general Australian students achieve comparably to other countries in mathematics (Thomson et al. 2004, 2010; Thomson and De Bortoli 2008). However, when these results are analyzed in relation to geographical location, significant variations in student achievement emerge (Fig. 3.1).

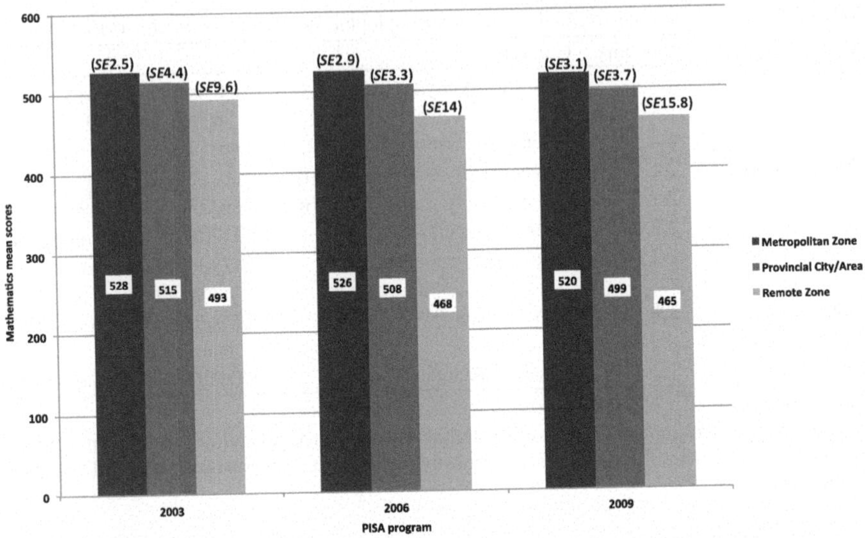

Fig. 3.1 Mean scores for student achievement for PISA 2003, 2006, and 2009 mathematics across geographical locations. *SE* standard error. (Data from Thomson et al. 2004, 2010; Thomson and De Bortoli 2008)

Reflecting upon Fig. 3.1 it is evident that in each of the PISA 2003, 2006, and 2009 testing periods, the mean score for student achievement in mathematics decreases with increasing distance from the Metropolitan Zone. While the variation across locations is apparent, note the larger decrease in the mean score for students in schools located in the Remote Zone between 2003 and 2006 with this score remaining reasonably level in 2009. The actual size of the gap in achievement is exemplified using a standard error (SE), which expresses variation about the mean. A lack of overlap between the SEs identifies major differences in the data. As published in the PISA reports, each of the differences highlighted here between geographical locations is statistically significant at a $p \leq 0.05$ level (Thomson et al. 2004, 2010; Thomson and De Bortoli 2008). Clearly, these data suggest that diversity across geographical location is evident.

A critical question raised given the Australian context is: *To what extent can these findings be explained by the low achievement of our Indigenous students?* As presented in Fig. 3.2, the mean scores for Indigenous and non-Indigenous students varied by 86 score points in PISA 2003, 80 in 2006, and 76 in 2009. According to official reports, these differences represent more than one proficiency level of mathematical literacy, which is the equivalent of between 2.5 years of schooling for 2003 and 2006, and 2 years of schooling for 2009 (De Bortoli and Thomson 2009; Thomson and De Bortoli 2008; Thomson et al. 2010).

So, while there is a significant difference between the achievement levels of Indigenous and non-Indigenous students, do these differences alone account for the variations in achievement across geographical location? Table 3.2 indicates that the proportion of Indigenous students participating in each PISA testing period

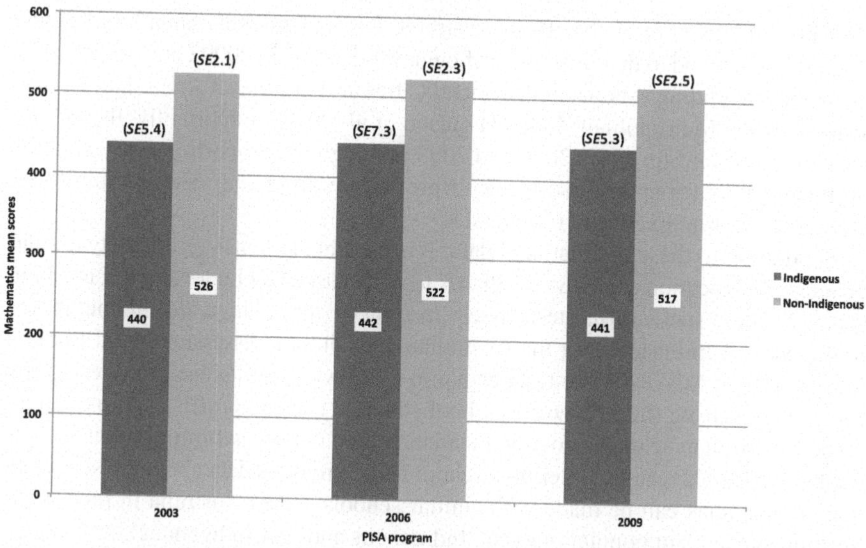

Fig. 3.2 Mean scores for Indigenous and non-Indigenous student achievement for PISA 2003, 2006, and 2009 mathematics. *SE* standard error, *PISA* Program for International Student Assessment. (Data from Thomson et al. 2004, 2010; Thomson and De Bortoli 2008)

Table 3.2 Proportions of Indigenous students participating in PISA. (Data from De Bortoli and Thomson 2009; Thomson et al. 2010)

	Total number of students included in PISA samples	Number of Indigenous students in PISA samples	Indigenous students in PISA samples (%)
PISA 2003	12,551	815	6.4
PISA 2006	14,170	1080	7.6
PISA 2009	14,251	1143	8

PISA Programme for International Student Assessment

represented between 6.4 and 8% of the total PISA student sample. According to Thomson and De Bortoli (2008) the Indigenous samples in PISA often present "an oversampling to reliably report results for this minority group" (p. 260).

The next point to clarify is the geographical location of the Indigenous samples for PISA 2003–2009 given that these data are readily available. In 2009, 9% of the Indigenous students involved in PISA were from Remote schools, with 19% in 2006 and 2% in 2003 (De Bortoli and Thomson 2009; Thomson and De Bortoli 2008; Thomson et al. 2010). Hence, the proportion of Indigenous students in relation to non-Indigenous students in Remote schools participating in each PISA testing period are simply insufficient to account fully for the consistently lower levels of achievement of students located in schools in the Remote Zone of Australia.

Having established these general patterns of variation across geographical locations, it is possible to go even further by considering these differences in relation to levels of proficiency. In terms of PISA, Level 2 is determined internationally as a baseline for student achievement while Levels 5 and 6 identify high-performing students (Thomson et al. 2010). In Australia, 33 % of students attending Remote Zone schools failed to demonstrate proficiency at Level 2 in 2009 compared to 19 % of students attending schools in Provincial Cities and Provincial Areas, and 15 % in schools in the Metropolitan Zone (Thomson et al. 2010). Importantly, these findings are consistent for PISA 2006 and 2003 although the proportions vary slightly. Sullivan elaborates upon these levels of proficiency in greater detail with a focus around classroom practice in Chapter 13.

In addition to these international data, it is possible to consider diversity around geographical location using the NAPLAN tests discussed by Leder and Lubienski in Chap. 2. These standardized tests assess reading, writing, language, and numeracy proficiency. Administered by the Australian Curriculum, Assessment and Reporting Authority (ACARA), the tests began in 2008 with results used to develop the *My School* website (http://www.myschool.edu.au). The site profiles approximately 10,000 Australian schools providing statistical and contextual information including the NAPLAN results. Entering a school name provides data about the school so that comparisons can be made with similar schools across Australia in relation to location and student populations (i.e., Indigenous and non-Indigenous).

To explore these data, Years 3 and 9 results from the most recently published 2012 NAPLAN tests for numeracy (ACARA 2012) were extracted and plotted. Figure 3.3 illustrates the findings for Year 3 students across geographical locations by comparing *All*, *non-Indigenous*, and *Indigenous* NAPLAN scores. The first pattern to observe is that the mean score for All students is highest in the Metropolitan Zone, then decreases by 21 points in Provincial City/Area, with a further drop of

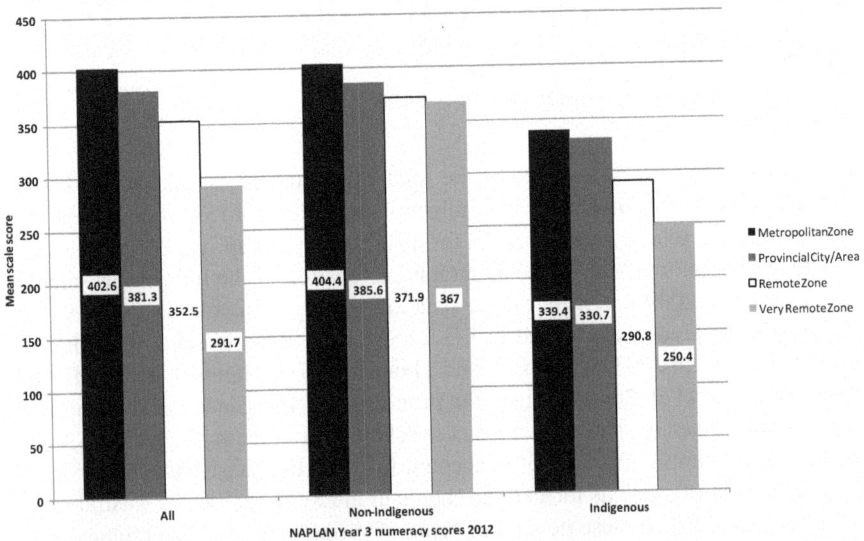

Fig. 3.3 NAPLAN Year 3 mean scores for 2012 for Indigenous and non-Indigenous students across geographical locations (Data from ACARA 2012). *NAPLAN* National Assessment Program Literacy and Numeracy. (Data from ACARA 2012)

29 points in the Remote Zone, and a major decrease of 61 points in Very Remote locations. As seen from the graph, the results for non-Indigenous students align with this pattern although the difference in mean scores between the various locations is less. Focusing on the Indigenous findings, the largest gaps occur between the Provincial City/Area and Remote Zone (i.e., 40 points), and the Remote Zone and Very Remote locations (40 points). Clearly, these results highlight the same diminishing lack of student achievement as the distance from the Metropolitan Zone increases as identified with the PISA data. However, unlike PISA, the ACARA report for 2012 does not provide enough information to be able to substantiate the level of statistical significance of these Year 3 NAPLAN data.

As a comparison, the NAPLAN Year 9 numeracy data (Fig. 3.4) present similar trends to those identified for Year 3 students with mean scores decreasing between geographical locations. However, what is noticeable immediately is that the gap in student achievement between the four locations is not as extreme as with the younger students. For example, there is a 21-point difference in the student data between the Metropolitan Zone and the Provincial City/Area locations, 20 points between Provincial City/Area and Remote schools, and 49 points between Remote Zone and Very Remote schools. Looking at the groups specifically, for non-Indigenous students the differences between these locations are 19, 6, and 9 points, respectively. In contrast, there is only a 7-point gap for Indigenous students between schools in the Metropolitan Zone and the Provincial City/Area but a 26-point gap between Provincial City/Area and Remote Zone schools, and a 29-point difference between Remote Zone and Very Remote schools. As mentioned earlier, while these gaps

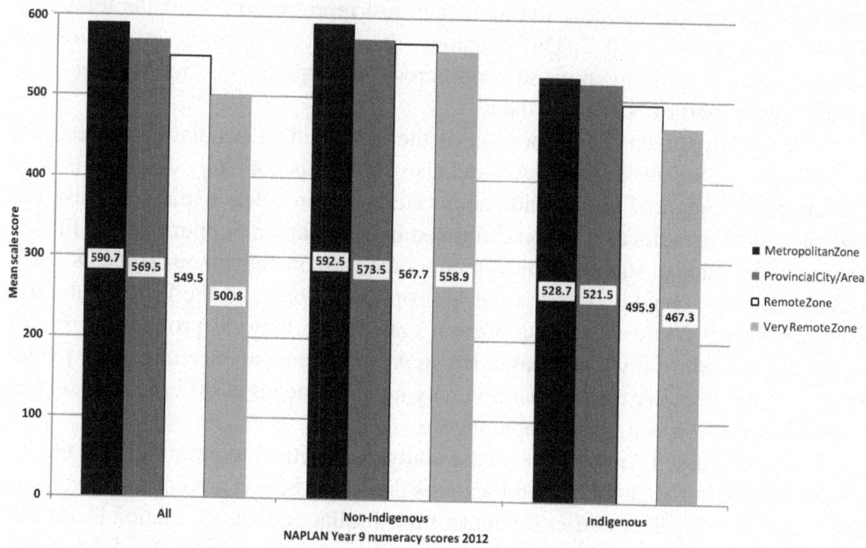

Fig. 3.4 NAPLAN Year 9 mean scores for 2012 for Indigenous and non-Indigenous students across geographical locations. *NAPLAN* National Assessment Program Literacy and Numeracy. (Data from ACARA 2012)

appear substantial in some cases, it is difficult to quantify the extent of these differences statistically without access to additional data.

As with PISA, in order to place these findings into perspective, it is necessary to consider the proportions of Indigenous and non-Indigenous students represented by the NAPLAN 2012 numeracy data. Table 3.3 provides a summary of the total

Table 3.3 Proportions of Indigenous students participating in NAPLAN 2011 numeracy as a proportion of total sample. (Data from ACARA 2012)

		NAPLAN participants for numeracy	Representativeness of the total population (%)
Year 3 (*N*=260,779)	Indigenous	12,374	88.2
	Non-indigenous	248,405	95.5
Year 9 (*N*=251,816)	Indigenous	10,112	75.8
	Non-indigenous	241,704	92.4

NAPLAN National Assessment Program Literacy and Numeracy

student population completing the numeracy test in Years 3 and 9 along with a break down of the number of Indigenous and non-Indigenous students in each cohort. The representativeness of each of these samples of the total population of Indigenous and non-Indigenous students is also included.

The Indigenous student sample included in both years is representative of the total number of Indigenous Year 3 students (i.e., 88.2 %) for the cohort across Australia and Year 9 students (i.e., 75.6 %). Having clarified this point, the proportion of Year 3 Indigenous students in NAPLAN 2012 represented only 4.7 % of the *total population* of Year 3 students who actually participated. Similarly, the Year 9 Indigenous students completing the numeracy test represented 4 % of the total Year 9 population tested in 2012. Unfortunately, data in relation to the proportions of Indigenous and non-Indigenous students across geographical location were not provided in the report of NAPLAN data.

Subsequently, given the proportion of the total student population represented by Indigenous students in NAPLAN and also PISA, it is not likely that their lack of achievement compared to non-Indigenous students can solely explain the discrepancy across geographical location discussed in this chapter. Supporting this further are the major gaps in student achievement noted for non-Indigenous students when mean scores are compared across the Metropolitan Zone and the Provincial City/Area locations. While these data are not conclusive, they do provide an insight about the breadth of diversity that exists in Australia and so encourages the reader to think more laterally about what diversity actually means at different scales (i.e., a macroscopic versus a microscopic view).

Having discussed Australia as a case study, it is critical to point out that the diversity in relation to geographical locations described here is not common with the only other developed countries demonstrating similar patterns in relation to student achievement being Canada, New Zealand, and Korea. In contrast, the USA experiences the opposite effect with students attending rural schools achieving more highly than students in urban or metropolitan schools (D'Amico and Nelson 2000; Barley and Beesley 2007).

Factors Impacting Diversity Across Location

The PISA and NAPLAN data are useful in providing a generalizable pattern in the student population but they only tell part of the story. For example, *why is it that Australian students in schools located in Provincial City and Provincial Areas achieve a lower mean score in PISA than their peers in schools in Metropolitan Zones?*

A pivotal factor identified in the research literature is socioeconomic status (SES), which can be considered in relation to student SES and school SES. The component most commonly explored in the literature and referenced in the media is student SES, which is determined using a number of criteria including parental education, occupation, and home address. It is represented in quartiles ranging from lowest to highest SES. These data are collected during PISA testing using demographic information from a separate student attitudinal survey completed at the same time. According to Australian PISA data for 2003–2009, as a student's SES increases from the lowest quartile to the highest quartile, mathematical achievement improves. For example, in PISA 2009, a 90-point difference was identified for mathematical literacy between students in the lowest and highest SES quartiles representing more than 2 years of schooling and more than one PISA proficiency level (Thomson et al. 2010). Similar results emerged for PISA 2006 and 2003 (see Thomson et al. 2004; Thomson and De Bortoli 2008).

McConney and Perry (2010) went one step further in their secondary analysis of Australian PISA 2006 by exploring the impact of school SES on student achievement. In their analysis, school SES was determined by averaging the student SES scores for individuals (within a school) who participated in PISA 2006. In brief they found that an increase in school SES was associated with an increase in student academic performance with this relationship holding regardless of the SES of the individual student. As an example, a student from a low SES background who achieved a mean score of 458 for PISA 2006 in a school with low SES would likely achieve a mean score of 533 in a school with high SES. These findings demonstrate the critical impact of school SES for which much less is known in the literature. The key point for this chapter is that when considering the factors impacting student achievement in Provincial City, Provincial Area, and Remote Zone schools, it is critical to be able to disregard student and also school SES in any statistical analysis because this becomes an overriding and confounding variable (Williams 2005).

In addition to SES, the literature highlights a number of other key factors in relation to geographical location including teacher attraction and retention, teaching resources, and professional isolation (Cresswell and Underwood 2004; Vinson 2002). To date, these factors have been subject to scrutiny in Australia over the last two decades resulting in a number of major reports including: Commonwealth Schools Commission (CSC 1988); the Human Rights and Equal Opportunity Commission (HREOC 2000); MCEETYA Task Force (2001); National Board of Employment, Education and Training (NBEET 1991); Ramsey (2000); and Roberts (2005). While these documents provide generic information, there is little known about the degree to which these factors potentially impact the teaching of mathematics, science, and information and communication technology (ICT) in schools across Australia.

To address this gap in the rural education literature, the National Centre for Science, Information and Communication Technologies and Mathematics Education for Rural and Regional (SiMERR) Australia conducted a national survey of teachers to explore the issues specifically (Lyons et al. 2006). In brief, five surveys were devised and distributed to collect data from secondary teachers of mathematics, science, and ICT; primary teachers; and parents or caregivers. While all teachers in schools located in the Provincial City/Area and the Remote Zone were invited to participate in the surveys, a representative sample of 20 % of teachers from the Metropolitan Zone was included to facilitate comparisons. In this section, some of the findings from the 547 secondary mathematics teachers from catholic, government, and independent schools who completed the survey (see Table 3.4) are discussed, along with other relevant literature.

Each teacher survey included items to extract demographic information and perceptions around the difficulties of attracting and retaining qualified staff; the degree of teacher access to professional development; student accessibility to learning opportunities; and the extent to which composite classes were implemented in schools. Surveys invited responses using multiple-choice formats, Likert-type rating scales, and open questions, with analyses conducted by a statistician with expertise in educational research. As part of the statistical analyses, school size and the SES of the school's location were controlled to minimize the confounding effects of these variables on the findings (Howley 2003; Lyons et al. 2006; McConney and Perry 2010). In presenting some of the findings from the survey, the four major SGLC categories identified in Table 3.1 are used. Additionally, details of the analyses have been kept to a minimum given that they are readily available from Lyons et al. (2006).

Table 3.4 Secondary mathematics teacher respondents by MSGLC category

	Main MSGLC categories				
	Metropolitan zone	Provincial city	Provincial area	Remote zone	
Criteria	Major cities pop. ≥ 100,000	Cities with pop. 25,000–99,999	Pop. < 25,000 and ARIA[a] plus score ≤ 5.92	Pop. < 25,000 and ARIA[a] plus score > 5.92	Total
Number of respondents mathematics (%)	142 (26 %)	132 (24.1 %)	240 (43.9 %)	33 (6 %)	547 (100 %)
Total teacher respondents to ALL surveys	580	661	1425	274	2940

[a] ARIA comprises a number of criteria, such as the physical road distance to the nearest service centre and the proximity to basic services (e.g., hospital. Data from Lyons et al. 2006). *ARIA* Accessibility and Remoteness Index of Australia, *MSGLC* Ministerial Council on Education, Employment, Training and Youth Affairs Schools Geographical Location Classifications

Attraction and Retention of Qualified Mathematics Teachers

A key issue for Australian schools in the Remote Zone is staff stability (Roberts 2005). To explore this aspect, teachers were asked to select the percentage of teachers leaving their school each year from 0 to 10%, 11 to 20%, 21%, and above. While each group of teachers reported a similar rate of change of staff for the 11–20% category, teachers in Metropolitan, Provincial City and Provincial Area schools reported their highest turnover of staff for the 0–10% category. Alternatively, the highest category for teachers in Remote Zone schools was >20% with a 43% response rate compared to 7% of teachers in Metropolitan Zone schools and 12% of teachers in Provincial Area schools. These findings were highly significant $p<0.001$ using a Chi-square test. As such, these results demonstrate that schools in the Remote Zone experience the highest rate of staff instability, which is especially unsettling given the turnover in staff included senior leadership in the schools (i.e., principals and assistant principals) in addition to classroom teachers.

To explore the attraction and retention of staff further, teachers were asked to rate the degree of difficulty experienced in filling vacant secondary mathematics teaching positions from options ranging from *Very Difficult* to *Not Difficult*. The results provided in Table 3.5 identify two significant differences $p<0.001$ (using a Chi-square test), which are shaded. These differences emerged because more teachers than expected in Metropolitan Zone schools indicated that it was *Not Difficult* to fill mathematics positions (i.e., 26%) while significantly more teachers than expected in Remote Zone schools (i.e., 65%) identified that it was *Very Difficult* to

Table 3.5 Reported difficulty of filling vacant secondary mathematics positions in MSGLC categories. (Data from Lyons et al. 2006)

			MSGLC categories				
			Metropolitan zone	Provincial city	Provincial area	Remote zone	Overall
How difficult is it to fill vacant secondary mathematics teaching positions?	Not difficult	*Respondents*	*33*	*14*	*28*	*1*	*76*
		% in row	43.4	18.4	36.8	1.3	100.0
		% in column	25.6	11.9	12.1	3.2	14.9
	Somewhat difficult	*Respondents*	*40*	*44*	*56*	*3*	*143*
		% in row	28.0	30.8	39.2	2.1	100.0
		% in column	31.0	37.3	24.2	9.7	28.1
	Moderately difficult	*Respondents*	*38*	*31*	*69*	*7*	*145*
		% in row	26.2	21.4	47.6	4.8	100.0
		% in column	29.5	26.3	29.9	22.6	28.5
	Very difficult	*Respondents*	*18*	*29*	*78*	*20*	*145*
		% in row	12.4	20.0	53.8	13.8	100.0
		% in column	14.0	24.6	33.8	64.5	28.5

MSGLC Ministerial Council on Education, Employment, Training and Youth Affairs Schools Geographical Location Classifications attract mathematics teaching staff to their schools. Focusing on these *Very difficult* responses further, there is clearly a stark contrast between the 65 and 34 % response rates from teachers in Remote Zone and Provincial Area schools, respectively, when compared to the 14 % of Metropolitan Zone teachers who selected this option.

To explore the qualifications of mathematics staff further aspect across Australia, teachers were asked whether they were currently teaching subjects for which they were not officially qualified (i.e., not having appropriate tertiary qualifications). Results were significant ($p<0.001$) using a Chi-square test (Table 3.6).

Table 3.6 Percentage of mathematics teachers in MSGLC categories required to teach subjects for which they were not qualified. (Data from Lyons et al. 2006)

		MSGLC categories				
		Metropolitan	Provincial city	Provincial area	Remote zone	Total
Secondary mathematics teachers	*Respondents*	*17*	*24*	*75*	*16*	*132*
	N within MSGLC	139	126	238	32	535
	% within MSGLC	12.2	18.9	31.5	50.0	24.6

MSGLC Ministerial Council on Education, Employment, Training and Youth Affairs Schools Geographical Location Classifications

As observed in the table, compared to Metropolitan Zone teachers (i.e., 12.2 %) twice as many teachers in Provincial Area schools (i.e., 31.5 %) and four times as many teachers in Remote Zone schools (i.e., 50 %) were expected to teach outside their subject expertise. As such, the rural education literature identifying the difficulties faced by schools in attracting and retaining qualified staff, which in this instance is mathematics, is supported and quantified by these national findings.

Professional Development Opportunities

Providing teachers with opportunities to engage in professional development is important but becomes increasingly challenging for teachers in locations outside of the Metropolitan Zone who must often travel to cities or larger centres to engage in these experiences (Roberts 2005). To explore this facet in the national survey, a range of professional development opportunities were provided with teachers responding regarding the *Importance* and *Availability* of these opportunities on Likert scales. By combining the two Likert scale ratings using a formula, an *Unmet Need* score was created with higher values indicating a greater unmet need for professional development (see Lyons et al. 2006). A summary of these findings is provided in Table 3.7.

The top four areas of highest need for secondary mathematics teachers included professional development opportunities for teaching higher order thinking, classroom management and organization, alternative teaching methods, and release from face-to-face teaching for collaborative activities. While these components were

Table 3.7 Mean "Need" scores[a] and standard deviations for teachers' ratings around professional development

Professional development items	Unmet need mean	SD
Professional development opportunities: teaching of higher-order skills	10.70	3.91
Professional development opportunities: classroom management and organization	10.47	4.04
Professional development opportunities: alternative teaching methods	10.34	3.98
Release from face-to-face teaching for collaborative activities	10.33	4.25
Effective communication between education authorities and teachers	9.92	3.72
Professional development opportunities: teach mathematics to gifted/talented students	9.89	3.72
Professional development opportunities: integrating technology into mathematics lessons	9.89	3.85
Professional development opportunities: teaching mathematics to special needs students	9.77	3.96
Collaboration with mathematics teachers in other schools	9.65	3.61
Professional development opportunities: methods for using group teaching strategies	9.60	3.80
Opportunities for observing teaching techniques of colleagues	9.49	3.97
Workshops to develop your ICT skills	9.47	3.82
Involvement in region/state-wide syllabus development/research projects	9.29	3.90
Financial support to attend external in-services/conferences	9.04	4.00
Opportunities for mentoring new staff	8.90	3.68
Opportunities to attend external in-services/conferences related to T&L mathematics	8.76	3.57
Professional development opportunities: use of graphics calculators	8.75	3.82
Professional development opportunities: outcomes/standards-based teaching	8.72	3.87
Opportunities to mark/mod external mathematics assessments	8.62	3.99
Professional development opportunities: teaching mathematics to Indigenous students	8.40	4.31
Professional development opportunities teaching mathematics to NESB students	8.29	3.99
Collaboration between mathematics teachers in your school	7.86	3.44

SD standard deviation, *ICT* information and communication technologies, *NESB* non-English-speaking background

[a] Items arranged in descending order of mean "need" score between 1 and 20. Adapted from Lyons et al. 2006

explored further (using Principal Components analysis and multivariate analysis of covariances, MANCOVAs) no significant differences emerged in relation to geographical location. Importantly, the lack of significance was due to the consistency with which teachers across geographical locations rated these items. As demonstrated in the following quotes, even teachers not located in Remote Zone schools identified issues associated with accessing professional development.

> A lot of professional development is available, but at great expense due to distance. It may involve large travel and accommodation cost, and/or extended time away from family. It is very hard to find help with the classroom (Provincial Area, New South Wales).

What PD? The school won't pay for airfares and nearly all PD is in Brisbane. Drive for hours and risk fatigue and accident, or don't go. Schools in regional areas should get bigger PD budgets as almost all good PD is in Brisbane (Provincial City, Queensland).

Student Learning Experiences

The *Shape of the Australian Curriculum: Mathematics* is the briefing paper that guided the development and writing of the Australian Curriculum: Mathematics. It reinforces the importance of engaging students in mathematics for two reasons. The first is the need to create "numerate citizens" (ACARA 2011, p. 10). The second is to encourage and support students to continue within the "mathematical pipeline" and not be filtered out too early (as occurs with streaming) thereby limiting their future career opportunities.

To explore this area of learner experience, teachers in the national survey rated a range of items as described earlier. The areas of greatest overall need (see Table 3.8) for mathematics teachers included students having opportunities to visit mathematics-related educational sites, being able to access alternative or extension activities in mathematics teaching programs for gifted and talented and for special needs students.

Further analysis of these data using a Principal Components analysis and MANCOVAs identified no significant differences indicating a high level of consistency in teacher responses across geographical locations. Hence, these results recognize the need for mathematics teachers regardless of location to be able to develop and implement tasks and activities in mathematics that both support but also challenge students' understanding of mathematics (Sullivan et al. 2013).

Table 3.8 Mean "Need" scores[a] and standard deviations for teachers' ratings of the student learning experience

Student learning need items	Unmet need mean	SD
Opportunities for students to visit mathematics related educational sites	9.36	3.70
Alternative/extension activities in mathematics teaching programs for gifted and talented students	9.22	3.58
Alternative/extension activities in mathematics teaching programs for special needs students	8.86	3.64
Alternative/extension activities in mathematics teaching programs for Indigenous students	8.47	4.16
Alternative/extension activities in mathematics teaching programs for NESB students	8.43	4.05
Teachers qualified to teach the mathematics courses offered in your school	8.15	3.06
Having the total indicative hours allocated to face-to-face teaching	8.12	3.48
Having the full range of senior mathematics courses available in your school	7.14	3.24
Student participation in external mathematics competitions and activities	5.92	2.49

SD standard deviation, *NESB* non-English-speaking background

[a] Items are arranged in descending order of mean "need" score between 1 and 20. Data source from Lyons et al. 2006

Composite Classes

There is considerable research around the positive and negative impacts of grouping students of different ages (i.e., multi-grade or composite classes) for teaching. While this approach is a usual practice in primary schools worldwide (Wilson 2003), composite classes have become more commonplace in secondary schools in Australia in response to financial considerations and the limited teacher expertise available in some subjects (e.g., senior mathematics and physics) (Roberts 2005). For example, allocating a secondary teacher to a class of four Year 11 mathematics students is not financially sustainable but if these students are merged with three Year 12 mathematics students then the class becomes viable. The alternative in some schools is not to offer senior mathematics or to enrol students with distance education providers for mathematics.

To explore this aspect and obtain specific information around secondary mathematics, teachers were asked in the national survey to respond to the following question: *Are some senior mathematics courses taught in composite classes in your school*? A summary of results is presented in Table 3.9.

These results demonstrate that composite classes for senior mathematics courses are especially prevalent in schools in the Remote Zone (66 %) but also in Provincial Areas (35 %). Some of the reasons identified by teachers in their comments about composite classes cited in Lyons et al. (2006, p. 131) included:

> The loss of specialist mathematics teachers results in teachers teaching out of their subject area and teaching composite classes (Provincial Area, NSW).
>
> It is a significant compromise for student learning to have composite classes in senior subjects like mathematics. To be successful, composite classes require students with a high degree of self-motivation, and independent learning skills. Many students in this school are from disadvantaged homes … Because the school has a small population, the more capable, and talented students are few in number, and have a significant pressure on them to fit the mould of under-performing (Provincial Area, NSW).
>
> Changes to syllabus requirements then impose great strain upon the mathematics teacher who is trying to cope with two different year levels (Provincial Area, Qld).

Composite classes and a number of associated interrelated factors as they relate to senior secondary classes are discussed in the next section of this chapter.

Table 3.9 Mathematics teachers in MSGLC categories reporting senior course taught in composite classes. (Data from Lyons et al. 2006)

		MSGLC categories			
		Metropolitan	Provincial city	Provincial area	Remote zone
Secondary mathematics teachers	Respondents $N=535$	139	127	237	32
	No (%)	93	83	65	34
	Yes (%)	7	17	35	66

MSGLC Ministerial Council on Education, Employment, Training and Youth Affairs Schools Geographical Location Classifications

Overall, the major findings emerging from the national survey indicate that secondary mathematics teachers in Remote Zone schools and to a lesser extent Provincial Area schools are more likely to experience greater staff instability (including senior leadership), share their teaching with staff not formally qualified to teach mathematics, and are more likely to teach composite classes for senior mathematics students. In contrast, it was interesting that a high degree of consistency occurred across geographical location in relation to teachers' ratings of their areas of need for professional development and in terms of their highest priorities for student learning experiences in mathematics.

Enhancing Greater Inclusivity

Given the diversity in student achievement discussed in this chapter for Australia, what can educators do to make a difference for students in Provincial Area and Remote Zone schools? How do we ensure greater inclusivity and thereby support students in continuing with mathematics into the senior secondary schooling? These questions are explored here from two perspectives: (1) by considering current practices in schools and the degree to which they may in fact hinder student engagement, achievement, and continuation in mathematics and (2) by thinking about ways of enhancing inclusivity for mathematics teachers in outer-Metropolitan schools.

School Practices

According to Stern (1994), many useful school practices including cooperative learning, multi-grade classrooms, and peer tutoring originated in rural schools where these strategies were developed with the purpose of enhancing the learning opportunities of students. So, on first appearance they appear to support inclusivity. However, the problem is that if not implemented or supported appropriately, these same practices may have the opposite effect. To exemplify this within the Australian context, many rural schools experience small class sizes for senior mathematics and physics. In order to offer the subject in the school, composite or multi-grade classes of Years 11 and 12 students are formed for collective teaching (Lyons et al. 2006). On one level this is positive in that it ensures that senior mathematics is still available in the school and provides an opportunity for multi-level inquiry and investigations given that Year 12 builds upon Year 11. The issue though is that it is the mathematics teacher who must carefully plan and coordinate lessons to ensure that both groups of students meet the specified curriculum requirements and demonstrate the mandatory learning outcomes. This is especially problematic with Year 12 where there is a credentialing examination (in all territories and states except Queensland) at the end of the year, with scores determining student acceptance into university courses.

Aligned to these composite classes is that there is often a reduction in the amount of time allocated for teaching. For example, although the mandatory time for Year

12 mathematics might be 5 hours per week and 5 hours for Year 11, the teacher may only be allocated 6 hours per week on the timetable for the composite class with the justification being that the class is small in number. Hence, instead of the teacher having the 9 mandatory hours, time is reduced to six requiring the mathematics teacher to juggle each lesson to ensure that both groups of students are covering the specified curriculum with reduced lesson time (Panizzon 2011).

While it might be argued that decreasing time in this manner is not equitable for students or teachers, it is possible that an experienced and highly qualified mathematics teacher might overcome the time issue ensuring quality practice and outcomes for the students. However, therein lies a confounding aspect in that many senior mathematics teachers in schools in the Provincial Area and Rural Zone in Australia are not necessarily qualified to teach senior mathematics nor are they experienced practitioners. In fact, most graduate and early career teachers find their first permanent full-time teaching positions in schools in these geographical areas (Lyons et al. 2006). So, in this time-pressured environment fuelled by a lack of teacher's mathematical expertise and experience is the reliance on "safe" pedagogical practices, such as teaching from the textbook through algorithms and extensive classroom practice (Pardhan and Mohammad 2005). Not surprisingly, teachers in these situations are less likely to deviate from those areas of mathematics where they have proficiency even if student interest is clearly evident.

A critical point to make here is that these practices (e.g., composite classes, reduced teaching time) and a lack of teacher expertise are equally relevant and applicable in some schools located in the Metropolitan Zone, particularly those in low SES areas (Calabrese-Barton 2007). The difference for teachers in Provincial Area and Remote Zone schools in Australia is that they are often dealing with a number of these practices simultaneously so that there is a compounding effect. While this is challenging in itself, a lack of access to pedagogical expertise and mentorship either within the school itself or from a learning community beyond the school (e.g., local teacher group) heightens the difficulties faced by these potentially "professionally isolated" mathematics teachers.

Importantly, this is not about blame but about creating greater awareness of the challenges faced in teaching in these contexts and the possible impacts of particular practices on student continuation in mathematics. It is only by recognizing the potential limitations that we can improve the learning opportunities and engagement of all students in mathematics.

Teacher Access to Professional Learning Communities

In considering inclusivity, there is often an immediate focus around what could be done differently for students, i.e., what pedagogies might teachers use? What mathematical tasks might build student confidence and experience in working mathematically to ensure continuation into the senior years (Sullivan et al. 2013)? How do we challenge our students to achieve their potential? However, if we are to address inclusivity in Provincial Area and Remote Zone schools, given the confounding factors and issues being juggled by these teachers, the emphasis needs to be more broadly targeted at teacher's professional growth and not just professional development. As highlighted in the research discussed in this chapter, access to professional networks is more likely in the Metropolitan Zone while problematic for teachers in Provincial City, Provincial Area, and Remote zone schools in Australia. Reasons for this include: (1) smaller populations of qualified and experienced mathematics teachers in these areas; (2) extensive distances between schools often restricting the frequency of face-to-face contact of teachers; (3) lower rates of teacher retention resulting in a high turnover of mathematics teachers; and (4) a higher proportion of graduate and early career mathematics teachers on staff (Lyons et al. 2006; Roberts 2005).

Critically, teachers require access to a professional learning community that is collegial, supportive yet challenging, comprising teachers with a range of experiences along with a shared interest in mathematics education (Bascia and Hargreaves 2000; Loughran 2010). In order to facilitate professional growth from graduate through to highly accomplished teacher as expected by the Australian Institute for Teaching and School Leadership (AITSL 2011), Westheimer (2008, p. 759) suggests that a learning community for teachers might focus upon six interrelated goals:

1. Improve teacher practice so students learn better
2. Make ideas matter to both teachers and students by creating a culture of intellectual inquiry
3. Develop teacher learning about leadership and school management
4. Promote teacher learning among novice teachers
5. Reduce alienation as a precondition for teacher learning
6. Pursue social justice and democracy

Yet, research indicates that establishing these networks is difficult given the tyranny of distance, while attempts to develop these electronically using technology tend to be hit-or-miss with teachers. Part of the reason for this lack of consistency is that initiating networks requires the building of relationships, which is always complex because it requires personal commitment, engagement, and a degree of negotiation by all involved (Corrigan 2004). The other issue is that once established, maintaining these networks requires an individual with the motivation, time, and technological expertise to coordinate the process. So here may be a place for the graduate teacher or early career teacher who may demonstrate greater technological knowledge and expertise along with a willingness to learn.

In recent years increasing access to Skype, Facebook, and other social media platforms (e.g., Twitter) have enhanced the opportunities for mathematics teachers to develop professional learning networks that move beyond the traditional face-to-face format. For example, in using Skype or Google hangouts, it is possible for small groups of teachers to talk face-to-face synchronously while sharing and discussing documents (e.g., mathematical tasks, assessment items). Critically, using technologies in this manner allows communities to overcome territory, state, and country boundaries to create a broader network of mathematics teachers.

The frustration, though, is that while governments espouse these kinds of technologies and their potential for enhancing educational opportunity, there is often no overarching framework or policy in place to ensure a unified and coordinated approach in establishing and maintaining these electronic possibilities into the future. Importantly, this involves more than the rollout of a fibre optic, wireless, and satellite infrastructure (e.g., the national broadband network currently being installed throughout Australia). It is about the need for models of best-practice and research evidence to inform how to use this infrastructure to ensure greater educational opportunity for teachers and students regardless of their geographical location.

Concluding Comments

Student diversity must be considered at a number of different levels if we are to meet the needs of students and facilitate greater access to mathematics over longer periods of time. Too often *catering for student diversity* becomes the role of the classroom teacher with little thought given to the wider possibilities and implications. As discussed in this chapter, access to large data sets, such as PISA and NAPLAN, provide the opportunity to consider student diversity in relation to geographical location. Using these data it appears that students located in schools in Metropolitan locations across Australia achieve significantly higher results for both PISA and NAPLAN when compared to students in all other geographical areas. Furthermore, as distance from these Metropolitan areas increases, there is a significant decrease in student achievement between Provincial Cities, Provincial Areas, Remote, and Very Remote areas. While prior generic research in rural education alluded to the possible inequity associated with geographical location in Australia, we now have the evidence as it relates to mathematics. Importantly though, in trying to understand these differences we must recognize the impact of SES and Indigeneity, which are confounding variables in Australia.

The purpose of this chapter is not to lay blame or to focus on the negative aspects around rural education but merely to *sow seeds* about the extent of diversity evident in Australia. Interestingly, this diversity exists not just in relation to students but also teachers with many of the outer-Metropolitan graduate teachers reflecting different needs to their more experienced colleagues positioned in Metropolitan schools. Ultimately, enhancing inclusivity in mathematics for students and teachers in these outer-Metropolitan locations requires a political imperative backed by a national framework that is supported financially over the long term. It must move beyond the classroom teacher and school into the wider community. Such directive is especially timely given the current financial commitments to increasing student engagement in mathematics and science in Australia and in many other countries.

References

Ainscow, M. (2007). Taking an inclusive turn. *Journal of Research in Special Educational Needs, 7*(1), 3–7.

Australian Bureau of Statistics (ABS). (2009). *Annual report 2009–2010*. Canberra: Australian Bureau of Statistics.

Australian Curriculum, Assessment and Reporting Authority (ACARA). (2011). The Australian curriculum: Mathematics version3.0. http://www.australiancurriculum.edu.au/Mathematics/Rationale. Accessed 10 April 2014.

Australian Curriculum, Assessment and Reporting Authority (ACARA). (2012). *NAPLAN achievement in reading, persuasive writing, language conventions and numeracy: National Report for 2012*. Sydney: ACARA.

Australian Institute for Teaching and School Leadership. (2011). National professional standards for teachers. Education services Australia for the ministerial council for education, early childhood development and youth affairs (MCEECDYA). http://www.teacherstandards.aitsl.edu.au/OrganisationStandards/Organisation. Accessed 3 Feb 2014.

Barley, Z. A., & Beesley, A. D. (2007). Rural school success: What can we learn? *Journal of Research in Rural Education, 22*(1), 1–16.

Bascia, N., & Hargreaves, A. (2000). Teaching and leading on the sharp edge of change. In N. Bascia & A. Hagreaves (Eds.), *The sharp edge of educational change, teaching, leading and the realities of reform* (pp. 3–26). New York: Routledge Falmer.

Calabrese-Barton, A. (2007). Science learning in urban settings. In S. Abell & N. Lederman (Eds.), *Handbook of research on science education* (pp. 319–343). Mahwah: Lawrence Erlbaum Associates.

Commonwealth Schools Commission. (1988). *Schooling in rural Australia*. Canberra: AGPS.

Corrigan, D. (2004). Understanding mentoring. In D. Corrigan & J. Loughran (Eds.), *Snapshots of mentoring: Vignettes of practice* (pp. 5–11). Clayton: Monash Print Services.

Cresswell, J., & Underwood, C. (2004). *Location, location, location: Implications of geographic situation on student performance in PISA 2000*. ACER research monograph number 58. Camberwell: ACER.

D'Amico, J. J., & Nelson, V. (2000). How on earth did you hear about us? A study of exemplary rural school practices in the Upper Midwest. *Journal of Research in Rural Education, 16*(3), 182–192.

De Bortoli, L., & Thomson, S. (2009). The achievement of Australia's indigenous students in PISA 2000–2006. http://research.acer.edu.au/ozpisa/6. Accessed 23 Sept 2013.

Howley, C. (2003). Mathematics achievement in rural schools. ERIC digest. http://permanent.access.gpo.gov/lps46365/lps46365/eric.ed.gov/PDFS/ED478348.pdf. Accessed 20 Nov 2013.

Human Rights and Equal Opportunities Commission (HREOC). (2000). Emerging themes: National inquiry into rural and remote education. http://www.humanrights.gov.au/pdf/human_rights/emerging_themes.pdf. Accessed 3 Feb 2014.

Jones, R. (2004). Geolocation questions and coding index. A technical report submitted to the MCEETYA performance measurement and reporting taskforce. http://www.mceecdya.edu.au/mceecdya/geolocation_questions_and_coding_index_11968.html. Accessed 23 Nov 2013.

Loughran, J. (2010). *What expert teachers do: Teachers' professional knowledge of classroom practice*. Sydney: Allen & Unwin, London: Routledge.

Lyons, T., Cooksey, R., Panizzon, D., Parnell, A., & Pegg, J. (2006). *Science, ICT and mathematics education in rural and regional Australia: Report from the SiMERR national survey*. Canberra: Department of Education, Science and Training.

McConney, A., & Perry, L. B. (2010). Science and mathematics achievement in Australia: The role of school socioeconomic composition in educational equity and effectiveness. *International Journal of Science and Mathematics Education, 8*, 429–452.

MCEETYA. (2008). Melbourne declaration on educational goals for young Australians. http://www.curriculum.edu.au/verve/_resources/National_Declaration_on_the_Educational_Goals_for_Young_Australians.pdf. Accessed 20 Nov 2013.

MCEETYA Task force on rural and remote education, training, employment and children's services. (2001). National framework for rural and remote education. http://www.mceecdya.edu.au/verve/_resources/rural_file.pdf. Accessed 7 July 2013.

National Board of Employment, Education and Training (NBEET). (1991). *Towards a national education and training strategy for rural Australia*. Canberra: AGPS.

Panizzon, D. (2011). Teaching secondary science in rural and remote schools: Exploring the critical role of a professional learning community. In D. Corrigan, J. Dillon, & R. Gunstone (Eds.), *The professional knowledge base of science teaching* (pp. 173–188). Dordrecht: Springer.

Pardhan, H., & Mohammad, R. F. (2005). Teaching science and mathematics for conceptual understanding? A rising issue. *Eurasia Journal of Mathematics, Science and Technology Education, 1*(1), 1–20.

Ramsey, G. (2000). *Quality matters: Revitalising teaching, critical times, critical choices*. Sydney: NSW Department of Education and Training.

Roberts, P. (2005). *Staffing an empty schoolhouse: Attracting and retaining teachers in rural, remote and isolated communities*. Sydney: NSW Teachers Federation.

Squires, D. (2003). Responding to isolation and educational disadvantage. *Education in Rural Australia, 13*(1), 24–40.

State of the Environment Advisory Council. (1996). *Australia: State of the environment*. Collingwood: CSIRO Publishing.

Stern, J. (1994). *The condition of education in rural schools*. Washington, DC: US Department of Education, Office of Educational Research and Improvement.

Sullivan, P., Clarke, D. M., & Clarke, B. A. (2013). *Teaching with tasks for effective mathematics learning*. New York: Springer.

Thomson, S., & De Bortoli, L. (2008). *Exploring scientific literacy: How Australia measures up*. Camberwell: Australian Council for Educational Research.

Thomson, S., Cresswell, J., & De Bortoli, L. (2004). *Facing the future: A focus on mathematical literacy among Australian 15-year-old students in PISA*. Camberwell: Australian Council for Educational Research.

Thomson, S., De Bortoli, L., Nicholas, M., Hillman, K., & Buckley, S. (2010). *Challenges for Australian education: Results from PISA 2009*. Camberwell: Australian Council for Educational Research. Accessed 2 April 2013.

United Nations Convention on the Rights of the Child. (1990). Declaration of the rights of the child. http://www.unicef.org/lac/spbarbados/Legal/global/General/declaration_child1959.pdf.

US Department of Education. (2001). *No child left behind*. Washington, DC: US Department of Education, Office of the Secretary.

Vinson, A. (2002). Inquiry into public education in New South Wales second report September 2002. http://www.nswtf.org.au/files/second_report.pdf. Accessed 19 Nov 2013.

Westheimer, J. (2008). Learning among colleagues. In M. Cochran-Smith, S. Feiman-Nemser & D. J. McIntyre (Eds.), *Handbook or research on teacher education: Enduring questions in changing contexts* (3rd ed., pp. 756–783). New York: Routledge.

Williams, J. H. (2005). Cross-national variations in rural mathematics achievement: A descriptive overview. *Journal of Research in Rural Education, 20*(5). http://www.jrre.psu.edu/articles/20-5.pdf. Accessed 7 July 2013.

Wilson, V. (2003). All in together? An overview of the literature on composite classes. SCRE Research Report 113. https://daedalus.lib.gla.ac.uk/bitstream/1905/171/1/113.pdf. Accessed 9 July 2013.

Chapter 4
Learners' Preferred Mathematical Task Types: The Values Perspective

Anastasios (Tasos) Barkatsas and Wee Tiong Seah

Introduction

A recent Nuffield Foundation review of more than 500 related published studies has found that in school mathematics, 'high attainment may be much more closely linked to cultural values than to specific mathematics teaching practices' (Askew et al. 2010, p. 12). These cultural values reflect what are considered important and worthy by teachers, students, parents, principals and the wider society with regards to effective mathematics teaching and learning. Being cognisant of what values are currently subscribed to and how certain values might be further emphasised may be the key components to improving mathematics teaching and learning effectiveness.

Significance of values and valuing in mathematics education may be understood in the context of values being regarded as a volitional variable (see Chapter 10, this volume). This perspective of valuing as an expression of intention, determination and will provides us with the opportunity to rethink mathematics learning and teaching in schools. In particular, in this chapter, we examine how students respond to the mathematical tasks they were presented in class. The absence of any pattern in student preferences among the types of mathematical tasks (see Sullivan 2010) can contribute to perceptions of diversities in the school mathematics classroom. Yet, as will be reported here, what students valued among the mathematical tasks they preferred were similar. As such, the values approach to interpreting students' preferences would be a more inclusive one than cognitive interpretations.

Student preferences among mathematical task types were examined with grade 5 and 6 students in Melbourne, Australia, and in Chongqing and Chengdu, China. In particular, students' personal preferences of a range of written mathematical tasks

A. (Tasos) Barkatsas (✉)
School of Education, Royal Melbourne Institute of Technology (RMIT) University, Melbourne, Australia
e-mail: tasos.barkatsas@rmit.edu.au

W. T. Seah
Faculty of Education, Monash University, Clayton, Australia
e-mail: WeeTiong.Seah@monash.edu

A. Bishop et al. (eds.), *Diversity in Mathematics Education*,
Mathematics Education Library, DOI 10.1007/978-3-319-05978-5_4,

would be evaluated to identify underlying values. The assumption here is that the students' preferences of mathematical task types reflect their own values as well.

We begin this chapter with a discussion of the construct of values as it pertains to mathematics education research, and how it is related to student preferences. The notion of mathematical tasks as it was adopted in the study presented here will also be outlined before the research context. Finally, the study data are interpreted given that they serve as an illustration of how we can interpret task preferences to identify underlying personal values.

Values

> Values refer to the principles, fundamental convictions, ideals, standards or life stances which act as general guides to behaviour or as points of reference in decision-making or the evaluation of beliefs or action and which are closely connected to personal integrity and personal identity (Halstead 1996, p. 5).

Values are not just embedded in teaching and learning (McLaren 1998), but are also intrinsic to school curriculum (Apple 2000). Indeed, according to Lovat and Clement (2008),

> values education [should] be at the heart of all pedagogical and curricular ventures and that any educational regime that sets out to exclude a values dimension in learning will be weakening its potential effects on all learning and student wellbeing, including academic learning. (p. 13)

Likewise, mathematics is a cultural product (Bishop 1988) and school mathematics is not value-free. According to Seah and Andersson (Chapter 10, p. 169).

> values are the convictions which an individual has internalised as being the things of importance and worth. What an individual values defines for her/him a window through which s/he views the world around her/him. Valuing provides the individual with the will and determination to maintain any course of action chosen in the learning and teaching of mathematics. They regulate the ways in which a learner's/teacher's cognitive skills and emotional dispositions are aligned to learning/teaching.

Bishop (1988) proposed initially that 'Western' mathematics is an expression of three complementary pairs of mathematical values: *rationalism* and *objectivism*, *control* and *progress*, and *openness* and *mystery*. However, in the mid-1990s, Bishop (1996) extended his idea of values in mathematics education beyond the discipline itself, to the learning context and the society within which the mathematics was situated. He proposed the categories of mathematical values, mathematics educational values (e.g. *neatness*) and general educational values (e.g. *honesty*). In the general learning context, Seah (2005) has demonstrated how the values subscribed to by principals, parents and students can be in conflict with those held by the teacher, as well as how these convictions were negotiated or co-valued as part of establishing the classroom norms and practices.

The development of a theoretical understanding of values and valuing (which we see above) has not meant that the identification of values is straightforward. Indeed, research by Clarkson et al. (2000) has highlighted that a direct means of gathering data (e.g. through interviews) might not be effective. The respondents themselves may not know the personal values embraced thereby affecting the extent to which these respondents are able to report or discuss particular values. Indeed, investigating the sociocultural nature of values calls for an innovative research approach to facilitate identification and analysis (Keitel 2003).

In the study presented here, values will be inferred from the preferences (Warren et al. 2011) of the individuals whose values we explored. Such an indirect way of identifying values in the mathematics learning context had been used over the last few years by Law et al. (2010), as well as Seah and Ho (2009). Similar to these other studies, the inferences would be made by researchers who were based in the same cultural setting as the respondents.

Mathematical Tasks

We regard a mathematical task as 'a classroom activity, the purpose of which is to focus students' attention on a particular mathematical idea' (Stein et al. 1996, p. 460). The quality of students' mathematics learning is moderated by the types of mathematical tasks they are exposed to (see Kilpatrick et al. 2001).

Various research studies have categorised mathematical tasks differently. For example, Swan's (2008) classification of task types is different from the approach adopted in the Quick And Simple And Reliable (QUASAR) study (see Stein et al. 2000). On the other hand, the Task Types in Mathematics Learning (TTML) project (Sullivan et al. 2009) examined teacher use of three particular types of mathematical tasks: *Type 1*, in which the tasks were designed to exemplify the mathematics through the use of models, representations, tools or explanations; *Type 2*, in which mathematics was situated within a contextualised practical situation; and *Type 3*, the open-ended tasks. One of the key findings of the TTML project was that each of the various task types is important to different learners at different times, and no one task type is redundant in the context of school mathematics education.

Yet, this diversity of student preferences has meant that 'the provision of meaningful and challenging mathematical tasks remains an issue in middle years' mathematics in Australia' (Clarke and Roche 2009, p. 722), contributing to the phenomenon of many students becoming disengaged with the subject in the secondary school years. Perhaps, one explanation might be that the ways in which tasks relate to student learning have not always been made explicit (Simon and Tzur 2004).

The study in this chapter contributes to our understanding in this regard by adopting the TTML task type classification. We argue that this task type classification is meaningful for the Chinese data too. While two of the three task types identified above were represented in the mainland Chinese curriculum, open-ended tasks (Type 3 in the TTML study) while present were new in the Chinese curriculum scene. This study affords us the opportunity to examine the extent to which the relatively newly introduced open-ended tasks have been embraced in the Chinese mathematics classroom. Indeed, by investigating the reasons for students' preferences of mathematical tasks, it is hoped that the underlying values might be inferred and teased out. In this way, the nature of these values, and any patterns observed, might provide a clue to the pedagogical features that best facilitate student learning and stimulate student preference, regardless of the task type. Identifying the pedagogical features valued by students enables us to shift our focus from the *form* of mathematical tasks to the *nature* of these tasks as we research how they might be better harnessed to bring about greater inclusivity of students in mathematics learning.

The Chinese data for this study were collected in two inland cities in South-western China, Chengdu and Chongqing. Both cities are significant economically and politically in that region of China, and also increasingly so nationally. Chengdu has a population of more than 14 million residents, and is the capital city of China's Sichuan province. It was selected as the site for the 2012 Fortune Global Forum. Three state primary schools in the urban city centre were randomly selected, and all grade 5 and 6 students in 15 classes across these schools were invited to participate. 1107 Chengdu students provided information via completing the questionnaire; 605 of these students were in grade 5 and 499 were grade 6 students.

Some 300 km away, Chongqing is a larger municipality with some 30 million residents. It is one of five national central cities in China (the others being Beijing, Shanghai, Tianjin and Guangzhou), and is an important industrial city in the Yangtze basin. The Chongqing data were collected from three randomly selected state primary schools. A similar number of students in these schools completed the questionnaire asking them what tasks they preferred and they learnt from the most. Overall, 609 students in grade 5 and 500 in grade 6 completed the questionnaire giving a total of 1109 students.

Importantly, regardless of the sample responses, these data are not representative of Chinese students' preferences and values in the same way that data collected in Beijing or Shanghai are not representative of Chinese norms and patterns. Mainland China's huge landmass (it has the third largest land area in the world, after Russia and Canada) with its diverse climate and terrain features has meant that no two mathematics lessons are taught in the same way even with a centralised, national curriculum. In this context, traditional norms and practices (including Confucianism) continue to play a key role in shaping pedagogical decisions and actions in the Chinese classroom (Ryan et al. 2009). According to the second author's many years' experience in schools across mainland China, the characteristics of typical Chinese-style mathematics lessons remain largely unchanged and include large classes, much teacher talk, regular homework and frequent testing.

The TTML task classifications apply to the Chinese students' learning experience in the classroom. While Task Types 1 (modelling) and 2 (contextualised) have featured in the Chinese mathematics curriculum for some time, Task Type 3 (open-ended) were introduced into schools (and in new editions of textbooks) throughout China in the latest, 2001 Basic Education Curriculum Reform exercise. The ways in which these tasks are posed to students in mainland China are no different from those we are accustomed to in the 'West'—teacher-posed verbal questions, as well as mathematical tasks which they are expected to answer individually, with a peer, or as part of a group. However, the Chinese students can also expect to be assigned mathematical questions to work on for homework. Another difference between Chinese and Australian primary school students' experiences with mathematical tasks would be the students' personal possession of an assigned mathematics textbook in the Chinese classroom, and its relative absence in Australia at the primary school level.

The Australian data were sourced across the state of Victoria, whose capital city Melbourne is the second largest city in the country. About 4.1 million of the state's 5.6 million residents live in Melbourne and its suburbs. Melbourne's multi-ethnic population is a distinguishing feature when considered against the Chinese cities, and should be kept in mind when we unpack the implications of the findings of this study. The student participants were from State, Catholic and Independent primary schools across urban and rural areas of the state. Of the 934 students who completed the questionnaire, 302 were in grade 5 and 392 in grade 6. The remaining students in the Victorian sample were in Years 7 and 8.

Research Design

This chapter reports on the quantitative phase of our sequential mixed methods research (Creswell 2009), which aims to map the field relating to the preference for and the use of different mathematical task types in Australian and Chinese primary school classrooms. The assumption adopted has been that an understanding of the types of mathematical tasks students prefer, and those from which they learn most, might shed light on how students' participation in mathematics learning can be made more inclusive.

The research method adopted was a questionnaire, translated from the one constructed for the TTML project for the same purpose. This 15-item questionnaire has a mix of Likert-scale items, ranking exercises and open-ended questions. In translating the questionnaire into the Chinese language, the contextual information of several items in the TTML version was modified to accommodate the societal realities in mainland China (see Seah et al. 2010).

We seek to report here the findings relevant to the following research question:

- What types of mathematical tasks are preferred by grade 5 and 6 students in Chongqing and Chengdu, China and in Victoria, Australia?

The data addressing this research question relate to questionnaire items 9 and 11, as shown in Appendix 1. In particular, item 9 is concerned with the mathematical topic area of number, whereas item 11 relates to the area of geometry.

Table 4.1 Friedman test results for student rank ordering of items 9ai–iii

Item	Mean rank		
	Chongqing	Chengdu	AUS
Task type 1 (9aiii)	2.09	1.80	1.82
Task type 2 (9ai)	1.76	2.17	2.44
Task type 3 (9aii)	2.14	2.02	1.73

Table 4.2 Friedman test results for student rank ordering of items 11ai–iii

Item	Mean rank		
	Chongqing	Chengdu	AUS
Task type 1 (11ai)	2.26	2.19	1.97
Task type 2 (11aiii)	1.75	1.93	2.15
Task type 3 (11aii)	1.99	1.88	1.88

Results and Discussion

Tables 4.1 and 4.2 summarise the students' mean rank of their preferred task types. Each of the six mathematics questions across the two questionnaire items has been tagged as Task Types 1, 2 or 3, in the same way that these were used in the TTML study (see above). Friedman tests were used to test for statistically significant differences in the ways students rank-ordered the three types of mathematical tasks (items 9ai–iii and 11ai–iii).

The differences in rankings were statistically significant for each student group:

- *Chongqing students*: [χ^2 (2, 1001) = 97.45, p < 7.45],
- *Chengdu students*: [χ^2 (2, 721) = 57.46, p < 7.46],
- *Australian students*: [χ^2 (2, 688) = 207.75, p < 07.75],

Thus, in the mathematical topic area of number (item 9), grade 5 and 6 students in Chongqing, China, preferred mathematical tasks in the order of Types 2 (contextualised tasks), 1 (modelling tasks) and 3 (open-ended tasks). Students in Chengdu, China preferred the tasks in the order of Types 1, 3 and 2, whereas their peers in Australia preferred number task Types in the order of 3, 1 and 2.

The differences in rankings were statistically significant for each student group:

- *Chongqing students*: [χ^2 (2, 1058) = 153.44, p < 0.001]
- *Chengdu students*: [χ^2 (2, 721) = 43.96, p < 0.001]
- *Australian students*: [χ^2 (2, 689) = 25.43, p < 0.001].

Thus, in the area of geometry (item 11), grade 5 and 6 students in Chongqing, China, preferred mathematical tasks in the order of Types 2, 3 and 1, students in Chengdu, China, in the order of Types 3, 2 and 1, and their peers in Victoria, Australia in the order of 3, 1 and 2.

Respondents were also asked to provide a reason for nominating a particular question as their favourite. Each of the questionnaires was read by two research assistants who each coded the open-ended entries in response to the question: 'What is

Table 4.3 Codes for reasons cited by respondents in ranking exercise

1	Challenging (more complex, lots of steps/have to think/I learn something new/improve)
2	Easy to do/understand (instructions clear)/I'm good at this/we do this a lot
3	Real life scenario
4	Involves a model/drawing/grid
5	Multiple solution strategies available, need to devise own strategies
6	Has more than one possible answer
7	Fun/I like this type of operation (e.g. division) or topic (e.g. area)
8	Numbers not words
9	Other

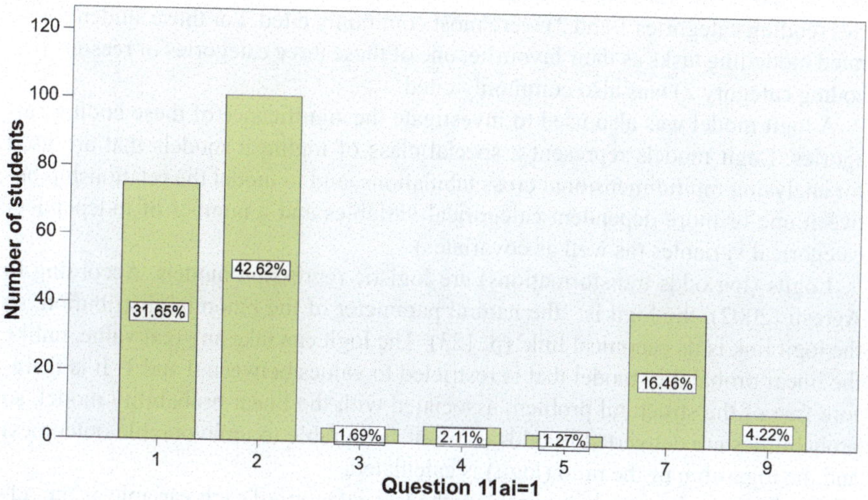

Fig. 4.1 Percentages by coding category for ranking item 11ai as favourite

regarded by the respondent as being important here'? The research assistants were local research students who understood the values concept who would also elicit the same worldview and values as the respondents. Indeed, the research assistants looked beyond the nominated reasons, to agree on the underlying values that were apparently held by the students. Mutual agreements of the value codes need to be achieved for each questionnaire response. Responses with similar value codes were then grouped together, thus ending up with nine categories of value codes listed in Table 4.3.

The bar chart in Fig. 4.1 displays the percentage in each coding category (1–9) of student respondents who nominated item 11ai (11ai = 1) over items 11aii and 11aiii as their favourite. Thus, for example, among the respondents who nominated item 11ai as the favourite over the other two items, 31.65 % of them did so because they valued the challenge (i.e. code category 1) that is inherent in that item. Bar charts for the other five questions may be similarly constructed.

As demonstrated in Fig. 4.1, most of the students who ranked modelling (Type 1) tasks as their favourite geometry tasks appeared to value them for one of the following three categories of reasons: *'Challenging (more complex, lots of steps/have to think/I learn something new/improve)'* (coding category 1, 31.65%), *'Easy to do/understand (instructions clear)/I'm good at this/we do this a lot'* (coding category 2, 42.62%) and *'Fun/I like this type of operation (e.g. division) or topic (e.g. area)'* (coding category 7, 16.46%). The same three categories of reasons were also most commonly cited by the other respondents who rated contextualised (Type 2) and open-ended (Type 3) tasks as favourite.

As for the number questions in item 9, these three categories of reasons were also most commonly cited by respondents who rated open-ended tasks as their favourite. Among those who rated contextualised tasks as their favourite, two of these categories (coding categories 1 and 2) were most commonly cited. For those students who rated modelling tasks as their favourite, one of these three categories of reasons (i.e. coding category 2) was also commonly cited.

A logit model was also used to investigate the significance of these coding categories. Logit models represent a special class of loglinear models that are used for analysing multidimensional cross tabulations, and to model the relationship between one or more dependent categorical variables and a number of independent categorical variables (as well as covariates).

Logits (log odds transformations) are logistic regression models. According to Agresti (2002), the logit is: 'the natural parameter of the binomial distribution, so the logit link is its canonical link' (p. 123). The logit can take any real value, unlike the linear probability model that is restricted to values between 0 and 1. It is therefore free of the structural problem associated with the linear probability model, so probabilities are converted to odds (ratio of favourable to unfavourable outcomes) and the logarithm of the ratio (logit) is calculated.

The SPSS logit procedure considers the last category of each variable as the reference category. So, the category 'fun' (coding category 7) is set to zero, and 9ai=3, 9aii=3, 9aiii=3, 10ai=3, 10aii=3, 10aiii=3, 11ai=3, 11aii=3, 11aiii=3, 12ai=3, 12aii=3 and 12aiii=3 (3 implies the task type students like the least) are all set to zero respectively in the corresponding logit models. The last two categories from Table 4.4 have not been used in the analysis because there were less than ten responses in each of these categories. The choice of 'fun/I like this type of operation' as reference category (against which the other reasons are compared) is credible, given that *fun* was one of the most popular of the students' values associated with effective mathematics learning (see Seah and Ho 2009). Tables 4.4 and 4.5 show the results of the logit analyses for the number and the geometry items.

The design for this test is governed by the following model: constant+q9ai+q9ai * q9b, constant+q9aii+q9aii * q9b, constant+q9aiii+q9aiii * q9b *(an identical model has been used for item 11),* where * represents $p<0.05$, **$p<0.01$; and ***$p<0.001$. The first number in each cell is the parameter estimate λ and the number in each parenthesis is e^{λ}. Three cell entries (representing the three student cohorts) from each

Table 4.4 Parameter estimates (λ, e^{λ}) summary for items 9i–9iii (number)

Reasons cited in ranking the items	Most preferred item 9ai (Number Type 2: Contextualised tasks)			Most preferred item 9aii (Number Type 3: Open-ended tasks)			Item 9aiii (Number Type 1: Modelling tasks			Least preferred items 9ai–9aiii
	Chongqing	Chengdu	AUS	Chongqing	Chengdu	AUS	Chongqing	Chengdu	AUS	
1: Challenging	-1.95*** (.14)	-.43 (.65)	-.64 (.53)	-1.25** (.29)	2.34 (10.38)	1.26*** (3.52)	-1.24*** (.29)	.19 (1.21)	-.38 (.68)	0 (1)
2: Easy to do	-.33*** (.72)	.67 (1.55)	-.02 (.98)	-1.8*** (.16)	1.8 (6.05)	-1.13** (.32)	1.79*** (5.99)	1.67 (5.31)	1.02*** (2.77)	0 (1)
3: Real life scenario	.81 (2.25)	2.71 (15.03)	1.22 (3.39)	id	-1.29 (.27)	.61 (1.84)	-1.27 (.28)	.66 (1.93)	1.22 (3.39)	0 (1)
4: Involves a model	id	id	id	1.79** (5.6)	-1.53 (.22)	id	1.64* (5.14)	1.04 (2.83)	id	0 (1)
5: Multiple solution strategies	id	.44 (1.55)	-.90 (.41)	id	id	id	-1.88 (.15)	.96 (2.61)	id	0 (1)
6: Has more than one possible answer	id		-2.51 (.08)	id	-3.13 (.04)	id	-.42 (.66)	1.04 (2.83)	id	0 (1)
7: Fun/I like this type of operation	0 (1)	0(1)	0 (1)	0 (1)	0 (1)	0 (1)	0 (1)	0 (1)	0 (1)	0 (1)

$^{*}p < .05$; $^{**}p < .01$; $^{***}p < .001$

Note: *id* means insufficient data for the calculation of the parameter estimates.

Table 4.5 Parameter estimates (λ, e^{λ}) summary for items 11i–11iii

Reasons cited in ranking the items	Most preferred Geometry type 1 modelling task (11ai)			Most preferred Geometry type 3 open-ended task (11aii)			Most preferred Geometry type 2 contextualised task (11aiii)			Least preferred items 9ai–9aiii
	Chongqing	*Chengdu*	*AUS*	*Chongqing*	*Chengdu*	*AUS*	*Chongqing*	*Chengdu*	*AUS*	0 (1)
1: Challenging	-.30 (.74)	2.41 (11.13)	1.18*** (3.25)	-.07 (.93)	1.11 (3.03)	-.17 (.84)	.31 (1.36)	.37 (1.51)	-1.30*** (.27)	0 (1)
2: Easy to do	-.683** (.50)	2.73 (15.33)	-.51 (.60)	.35 (1.37)	1.30 (3.67)	1.09***	.36 (1.4)	1.48 (4.39)	-.77* (.46)	0 (1)
3: Real life scenario	-.57 (.56)	.74 (2.10)	1.85* (6.36)	*id*	-.86 (.42)	*id*	.95 (2.56)	-1.90 (.15)	-1.30 (.27)	0 (1)
4: Involves a model	-.98* (.37)	.167 (1.18)	*id*	1.26 (3.5)	-1.24 (.29)	1.34** (3.82)	.008 (1.08)	-1.45 (.23)	-1.22** (.29)	0 (1)
5: Multiple solution strategies	-.86 (.42)	1.10 (3.00)	*id*	1.38 (4.0)	.43 (1.54)	*id*	-.99 (.37)	-.14 (.87)	*id*	0 (1)
6: Has more than one possible answer	*id*	-1.30 (.27)	*id*	*id*	-1.96 (.14)	*id*	*id*	-1.28 (.28)	*id*	0 (1)
7: Fun/I like this type of operation	0 (1)	0 (1)	0 (1)	0 (1)	0 (1)	0 (1)	0 (1)	0 (1)	0 (1)	0 (1)

*$p < .05$; **$p < .01$; ***$p < .001$

Note: *id* means insufficient data for the calculation of the parameter estimates.

of the Tables 4.5 and 4.6 will be discussed; the other cell entries may be similarly interpreted.

The parameter estimate for *real life scenario* being the favourite number contextualised tasks (item 9ai) for *Chongqing* grade 5 and 6 students is 0.81 (Table 4.4, code 3, highlighted second column, fifth row). The value of e^{λ} is $e^{0.081}=2.25$. That is, based on the model, the odds of grade 5 and 6 Chongqing students in the study nominating *real life scenario* as a reason for the contextualised task being a favourite over it being nominated when it is least liked, is more than twice the odds of nominating *fun/I like this operation* as a reason for the same task type being a favourite over it being nominated when it is least liked.

For *Chengdu* grade 5 and 6 students, the parameter estimate for *real life scenario* being the favourite for the same item (Table 4.4, code 3, highlighted third column, fifth row) is 2.71. The value of e^{λ} is $e^{2.71}=15.03$. That is, based on the model, the odds of grade 5 and 6 Chengdu students in the study nominating *real life scenario* as a reason for the contextualised task being a favourite over it being nominated when it is least liked, is 15 times the odds of nominating *fun/I like this operation* as a reason for the same task type being a favourite over it being nominated when it is least liked.

For Australian grade 5 and 6 students, the parameter estimate for *real life scenario* being the favourite for the same item (Table 4.4, code 3, highlighted fourth column, fifth row) is 1.22. The value of e^{λ} is $e^{1.22}=3.39$. That is, based on the model, the odds of grade 5 and 6 Australian students in the study nominating *real life scenario* as a reason for the contextualised task being a favourite over it being nominated when it is least liked, is more than three times the odds of nominating *fun/I like this operation* as a reason for the same task type being a favourite over it being nominated when it is least liked.

The parameter estimate for *'challenging'* being the favourite for item 11ai for *Chongqing* grade 5 and 6 students is −0.30 (Table 4.5, code 1, highlighted second column, third row). The value of e^{λ} is $e^{-0.30}=0.74$. That is, based on the model, the odds of grade 5 and 6 Chongqing students in the study nominating *challenging* as a reason for the modelling task being a favourite over it being nominated when it is least liked is 0.74 times the odds of nominating *fun/I like this operation* as a reason for the same task type being a favourite over it being nominated when it is least liked.

For *Chengdu* grade 5 and 6 students, the parameter estimate for the *challenging* task being the favourite for the same item is 2.41 (Table 4.5, code 1, highlighted third column, third row). The value of e^{λ} is $e^{2.41}=11.13$. That is, based on the model, the odds of grade 5 and 6 Chengdu students in the study nominating *'challenging'* as a reason for the modelling task being a favourite over it being nominated when it is least liked, is at least 11 times the odds of nominating *fun/I like this operation* as a reason for the same task type being a favourite over it being nominated when it is least liked.

For *Australian* grade 5 and 6 students, the parameter estimate for *'challenging'* being the favourite for the same item is 1.18 (Table 4.5, code 1, highlighted fourth column, third row). The value of e^{λ} is $e^{1.18}=3.25$. That is, based on the model, the odds of grade 5 and 6 Australian students in the study nominating *'challenging'* as a reason for the modelling task being a favourite over it being nominated when it is at least liked, is statistically significantly more than three times the odds of nominating *fun/I like this operation* as a reason for the same task type being a favourite over it being nominated when it is least liked.

For an explanation on how the parameter estimates shown in Tables 4.4 and 4.5 are calculated, please see Appendix 2 (Table 4.6).

Synoptically, it could be argued that there is an increased likelihood that grade 5 and 6 students' favourite task choices for number-type tasks were *challenging*, *easy to do* and *real life scenario* compared to *fun*. Some of the reasons cited in ranking the items for *challenging, easy to do, real life scenario* and *involves a model* were statistically significant (see Table 4.5 entries with one, two or three asterisks, which correspond to $*p<0.05$; $**p<0.01$; $***p<0.001$) respectively.

Similarly, for the geometry-type tasks, it could be argued that there is an increased likelihood that grade 5 and 6 students' favourite task choices (over the same task type being least liked) were relative to their valuing of fun, with differing strengths, *challenging, easy to do, real life scenario and involves a model,* compared to *fun.* Some of the reasons cited in ranking the items for these ranking categories were statistically significant (see Table 4.5 entries with one, two or three asterisks, which correspond to $*p<0.05$; $**p<0.01$; $***p<0.001$) respectively.

Concluding Remarks

The data presented in this chapter relate to grade 5 and 6 students' preferences among three types of mathematical tasks. These students were from the Chinese cities of Chongqing and Chengdu, and from the state of Victoria, Australia. It was found that for both number and geometry items, students in Australia preferred most to engage with tasks involving open-ended questions, followed by modelling tasks, and lastly, contextualised tasks.

In our Chinese data sets, however, there was in general no pattern of preference across the two cities by topic types or by question types. An exception is the Chongqing data, in which the student participants reported a preference for contextualised tasks, which was the least preferred task type among students in Australia. Otherwise, the Chinese students' preferences among the task types for both the number and geometry topics were without pattern.

There is thus a diversity of what students prefer among mathematical tasks. This is so, even for students in Chongqing and Chengdu, cities which are located merely some 300 km apart in a country with a centralised mathematics curriculum. Surely, the relative similarity of the cultures in Chongqing and Chengdu should lead to similar student preferences, something which was, however, not observed in our data.

Bearing in mind the objective of this research study, how do these results shed light on making students' participation in mathematics learning more inclusive, thereby optimising student achievement? Perhaps, as suggested by Sullivan (2010), teachers should make use of a variety of tasks in the classroom.

Alternatively, another way of enhancing learning and achievement that is not guided by task preference might be about rethinking the tasks themselves. In particular, among the task types which were the most preferred across the different sites, what common feature(s) ran through them? While a variety of reasons were provided for preferring particular task types, three of them emerged to be statistically significant relative to the valuing of *fun*. These were the valuing of *challenge, easiness* and *real life scenarios*. Remarkably, these values appeared to be independent of the mathematics topic type, and regardless of whether the students were in China or in Australia. The consistency with which students valued *challenge, easiness* and *real-life scenarios* across the large student samples generated in this study—a total of 3150 respondents across the three places—represents a strong message for efforts that facilitate inclusiveness for all students.

While some students valued *challenge* and not *easiness*, and some others valued *easiness* but not *challenge*, there is the possibility that there were students who valued both *challenge* and *easiness* in mathematical tasks. As contradictory as these two values appeared to be in relation to each other, it is possible that a student valued both of them, given the way the students' reasons were categorised. For example, one may value *easiness* for the many times one goes through in completing a mathematical task (see earlier listing of some reasons for coding category 2), which is also associated with the importance one places in learning something new (a reason categorised as the valuing of *challenging*, referring to the earlier listing of some reasons for coding category 1).

The students' valuing of *real life scenarios* is not surprising given that student disengagement with the subject has often been associated with questions of when they would ever get to apply or use particular mathematical concepts. 'When will we ever need to use this?' is a familiar question posed by students in mathematics classes. As such, mathematical tasks that provide authentic learning and application of concepts and skills—either through contextualised situations or open-ended ones—would be highly valued by students during their mathematics learning journey.

Thus, what we see here is that students' preference for specific types of mathematical tasks varies across different geographical locations and different mathematical topics. Such differences may well be explained by differences in cultures and in the nature of the mathematical content. However, an examination of the underlying student values that inform these personal preferences pointed to a remarkably consistent valuing of *challenge, easiness* and *real-life scenarios* across Australia and China, between the states of Chongqing and Chengdu within China, and across both number and geometry topics. A rethink of how the data is interpreted, focussing on investigating the values which guide personal preferences, has thus provided us with a different perspective to the phenomenon being researched. This view presents mathematics students from different education systems as a more inclusive group of learners. The values approach may then represent one productive way of facilitating and maintaining student engagement and learning in school mathematics.

Appendix 1

In this table there are four maths questions that are pretty much the same type of mathematics content asked in different ways.

We don't want you to work out the answers.

Put a 1 next to the type of question **you like to do most**, 2 next to the one you like next best, and 3 next to the type of question **you like least**:

9ai An adult cinema ticket costs RMB25, and a child ticket costs RMB12. How much would the tickets cost for 2 adults and 4 children to watch a movie?
9aii 2 adults and 4 children s pent RMB120 on movie tickets. How much might an adult ticket and a child ticket cost?
9aiii 25 X 2 + 12 X 4 =

You like to do this type of question (the one you put a 1 against) the most because: ______

______

This is the same table as in question 9. This time we want you to put a 1 next to the type of question that helps **you learn the most**, 2 next to the one you learn from second most and 3 next to the type **you learn the least** from:

10ai An adult cinema ticket costs RMB25, and a child ticket costs RMB12. How much would the tickets cost for 2 adults and 4 children to watch a movie?
10aii 2 adults and 4 children spent RMB120 on movie tickets. How much might an adult ticket and a child ticket cost?
10aiii 25 X 2 + 12 X 4 =

The reason this type of question (the one you put a 1 against) helps me learn the most is: ______

______

In this table there are four more maths questions that are pretty much the same type of mathematics content asked in different ways.

We don't want you to work out the answers.

Put a 1 next to the type of question **you like to do most**, 2 next to the one you like next best, and 3 next to the type of question **you like least**:

11ai Find the area of the following figure. 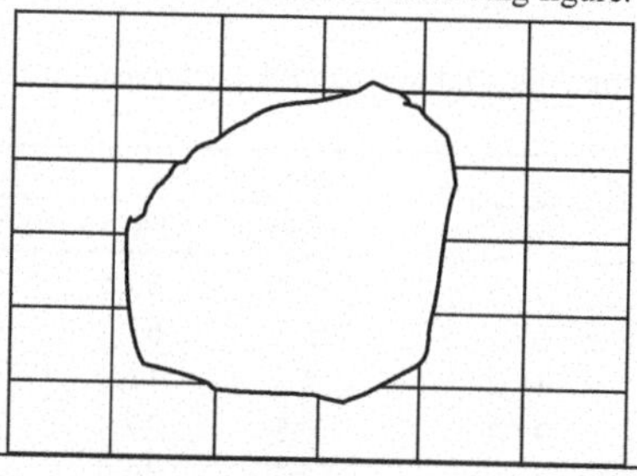	
11aii If the area of a figure is 10 square units, what might the shape of the figure be?	
11aiii An athletic track is made up of two straight sections and two semi-circles. The straight section is 100m long. What is the area of the athletic track? 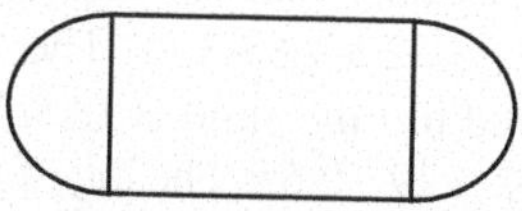	

You like to do this type of question (the one you put a 1 against) the most because:

This is the same table as in question 11. This time we want you to put a 1 next to thetype of question that **helps you learn the most**, 2 next to the next most helpful, 3 for theone after that, and 4 next to the type of question that **least helps you learn:**

12ai Find the area of the following figure. 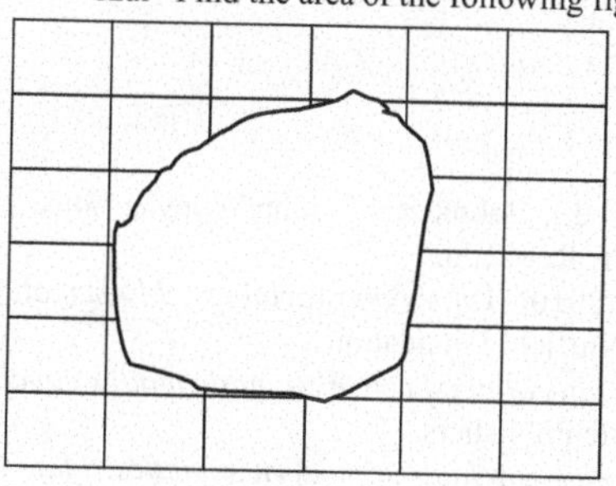	
12aii If the area of a figure is 10 square units, what might the shape of the figure be?	
12aiii An athletic track is made up of two straight sections and two semi-circles. The straight section is 100m long. What is the area of the athletic track? 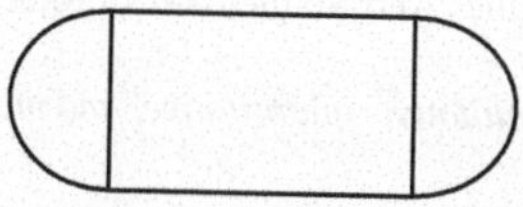	

The reason this type of question helps me learn the most is:

Appendix 2

Table 4.6 Indicative partial table segment of Chengdu students' ranking and task preferences used in the calculation of parameter estimates (Q9ai)

q9b (Ranking)	q9ai (Type 2 task)	Observed		Expected	
		Count	%	Count	%
1 (Challenging)	1 (Prefer most)	32.5	20.8	32.5	20.8
	2 (Prefer second most)	105.5	67.4	105.5	67.4
	3 (Prefer least)	18.5	11.8	18.5	11.8
7 (Fun)	1 (Prefer most)	9.5	37.3	9.5	37.3
	2 (Prefer second most)	12.5	49.0	12.5	49.0
	3 (Prefer least)	3.5	13.7	3.5	13.7

Each cell has 0.5 added to it, which is the SPSS default value

From the data shown in Table 4.6, the observed odds that 'Chengdu students prefer *Challenging* tasks most, compared to *Fun*' are 32.5/9.5=3.42. The 'Chengdu students prefer *Challenging* tasks least compared to *Fun*' ratio is 18.5/3.5=5.29. The ratio of these two odds, the *odds ratio* is 3.42/5.29=0.65. The log of the odds ratio, called the *log odds ratio*, is *Ln (0.65)*=−*0.43* (which is the top value highlighted in Table 4.6, second column, third row).

Note that the parameter estimates (log odds ratios) in SPSS (Tables 4.4 and 4.5) are logarithmic numbers and we have to take the exponential of the value in each cell in order to derive the odds ratios from the log odds ratios, i.e. in our case: $e^{-0.43}=0.65$ (which is the highlighted value in the bracket, Table 4.4, 3rd column, 2nd row).

References

Agresti, A. (2002). *Categorical data analysis* (Second ed.). Hoboken, NJ: John Wiley & Sons.

Apple, M. W. (2000). *Official knowledge* (2nd ed.). NY: Routledge.

Askew, M., Hodgen, J., Hossain, S., & Bretscher, N. (2010). *Values and variables: Mathematics education in high-performing countries*. London: Nuffield Foundation.

Bishop, A. J. (1988). *Mathematical enculturation: A cultural perspective on mathematics education*. Dordrecht, The Netherlands: Kluwer Academic Publishers.

Bishop, A. J. (1996, June 3-7). *How should mathematics teaching in modern societies relate to cultural values — some preliminary questions*. Paper presented at the Seventh Southeast Asian Conference on Mathematics Education, Hanoi, Vietnam.

Clarke, D., & Roche, A. (2009). Opportunities and challenges for teachers and students provided by tasks built around 'real' contexts. In R. Hunter, B. Bicknell, & T. Burgess (Eds.), *Crossing divides: Proceedings of the 32nd annual conference of the Mathematics Education Research Group of Australasia* (pp. 722–726). Palmerston North: MERGA.

Clarkson, P. C., Bishop, A. J., Seah, W. T., & FitzSimons, G. (2000, December 4-7). Methodology challenges and constraints in the Values And Mathematics Project. Paper presented at the AARE Sydney 2000, Sydney, Australia.

Creswell, J. W. (2009). *Research design: Qualitative, quantitative, and mixed methods approaches* (3rd ed.). Thousand Oaks: Sage.

Halstead, M. (1996). Values and values education in schools. In J. M. Halstead & M. J. Taylor (Eds.), *Values in education and education in values* (pp. 3-14). London: Falmer Press

Kilpatrick, J., Swafford, J., & Findell, B. (2001). *Adding it up: Helping children learn mathematics*. Washington, DC: National Academy.

Keitel, C. (2003). *Values in mathematics classroom practice: The students' perspective*. Paper presented at the Conference of the Learners' Perspective Study international research team, Melbourne, Australia.

Lovat, T., & Clement, N. (2008). Quality teaching and values education: Coalescing for effective learning. *Journal of Moral Education, 37*(1), 1–16.

Law, H. Y., Wong, N. Y., & Lee, N. Y. L. (2010). The third wave: Regional and cross-regional studies of values in effective mathematics education - Hong Kong. In Y. Shimizu, Y. Sekiguchi & K. Hino (Eds.), *Proceedings of the 5th East Asia Regional Conference on Mathematics Education* (Vol. 2, pp. 111-118). Tokyo, Japan: Japan Society of Mathematical Education

McLaren, P. (1998). *Life in schools: An introduction to critical pedagogy in the foundations of education*. NY: Longman.

Ryan, J., Kang, C., Mitchell, I., & Erickson, G. (2009). China's basic education reform: an account of an international collaborative research and development project. *Asia Pacific Journal of Education, 29*(4), 427-441.

Seah, W. T. (2005). Negotiating about perceived value differences in mathematics teaching: The case of immigrant teachers in Australia. In H. Chick & J. L. Vincent (Eds.), *Proceedings of the 29th conference of the International Group for the Psychology of Mathematics Education* (Vol. 4, pp. 145–152). Melbourne: PME.

Seah, W. T., Barkatsas, A., Sullivan, P., & Li, Z. (2010). Chinese students' perspectives of effective mathematics learning: An exploratory study. In T. Desmond (Ed.), *The Asian conference on education official conference proceedings 2010* (pp. 389–403). Japan: International Academic Forum.

Seah, W. T., & Ho, S. Y. (2009). Values operating in effective mathematics lessons in Australia and Singapore: Reflections of pre-service teachers. In M. Tzekaki, M. Kaldrimidou & H. Sakonidis (Eds.), *Proceedings of the 33rd conference of the International Group for the Psychology of Mathematics Education* (Vol. 5, pp. 57–64). Thessaloniki, Greece: International Group for the Psychology of Mathematics Education.

Simon, M. A., & Tzur, R. (2004). Explicating the role of mathematical tasks in conceptual learning: An elaboration of the hypothetical learning trajectory. *Mathematical Thinking and Learning, 6*(2), 91–104.

Stein, M. K., Grover, B. W., & Henningsen, M. A. (1996). Building student capacity for mathematical thinking and reasoning: An analysis of mathematical tasks used in reform classrooms. *American Educational Research Journal, 33*(2), 455–488.

Stein, M. K., Smith, M. S., Henningsen, M. A., & Silver, E. A. (2000). *Implementing Standards-based mathematics instruction: A casebook for professional development*. New York: Teachers College Press.

Sullivan, P. (2010). *Learning about selecting classroom tasks and structuring mathematics lessons from students teaching mathematics? Make it count: What research tells us about effective teaching and learning of mathematics* (pp. 53–55). Victoria: Australian Council for Educational Research.

Sullivan, P., Clarke, D. M., Clarke, B. A., & O'Shea, H. (2009). Exploring the relationship between tasks, teacher actions, and student learning. In M. Tzekaki, M. Kaldrimidou, & H. Sakonidis (Eds.), *Proceedings of the 33rd conference of the international group for the psychology of mathematics education* (Vol. 5, pp. 185–192). Thessaloniki: PME.

Swan, M. (2008). Designing a multiple representation learning experience in secondary algebra. *Educational Designer, 1*, 1–17.

Warren, C., McGraw, A. P., & Van Boven, L. (2011). Values and preferences: Defining preference construction. *Wiley Interdisciplinary Reviews: Cognitive Science*, 2, 193-205. doi: 10.1002/wcs.98

Chapter 5
Gender and Technology: A Case of Graphics Calculators in the Singaporean Mathematics Curriculum Context

Hazel Tan

Introduction

Equity and inclusivity issues in general mathematics education as well as relating to technology use in mathematics education have been highlighted in the Australian and other western education context. For example, in the statement for use of calculators and computers by the Australian Association of Mathematics Teachers (AAMT), there was a section on equity: "Every care must be taken to ensure that the use of technology does not contribute to increased in equity amongst individuals and groups already recognised as being disadvantaged by reason of race, gender, disability or socio-economic status" (AAMT 1996, Statement on use of calculators and computers for mathematics in Australian schools, p. 5).

Yet, the gender difference in mathematics education outcomes Australia still persists across the different year levels (see Chapter 2 by Leder and Lubienski; Forgasz and Tan 2010). Internationally, there are disparate findings in the learning outcomes, with gender differences widening in some countries such as Australia and narrowing in other countries such as Mexico Forgasz et al. (2010). This chapter focuses on the use of graphics calculators (GCs) in the senior secondary Singaporean mathematics curriculum. Research involving gender and technology use in mathematics in the context of an Asian country are relatively rare. Hence, it is hoped that this chapter would enrich the discussion and debate. In Singapore, the issue of equity, in particular equity related to gender, is very seldom explicitly mentioned in any of the curriculum or policy documents. Being one of the countries with consistently top performing scores in the Trends in International Mathematics and Science Study (TIMSS) series of international studies in 1995, 1999, 2003, 2007 and 2011 (Mullis et al. 2012) the small island state of Singapore and its educational system have since gained international interest. The Organisation for Economic Cooperation and Development (OECD) has highlighted Singapore's education system as high

H. Tan (✉)
Faculty of Education, Monash University, Clayton, Australia
e-mail: hazel.tan@monash.edu

A. Bishop et al. (eds.), *Diversity in Mathematics Education*,
Mathematics Education Library, DOI 10.1007/978-3-319-05978-5_5,

performing based in the wake of its high performance in the 2009 Programme for International Student Assessment (PISA, http://www.oecd.org/pisa/).

With regard to technology use, GCs have been allowed in the Singaporean preuniversity (equivalent to senior secondary or grades 11 and 12) examinations for all the mathematics subjects in since 2007 (Ministry of Education Singapore [MOE], 2007). However, not much research had been conducted in this area. Since entrance into tertiary education depends on students' preuniversity examination performance, the question is whether the use of technology (GCs in this case) widens or narrows the gender gap (if any).

In Chapter 7, Askew used Gutiérrez's (2007) definition of equity to be such that mathematical achievement should not be predictable on the basis of labels such as gender, indigenity and disability. In terms of school mathematics achievement, in Singapore there are no publicly available data on the mathematical performance of boys and girls, unlike in Australia or the USA where the National Assessment Program—Literacy and Numeracy (NAPLAN) and National Assessment of Educational Progress (NAEP) results are published (see Leder and Lubinski, Chapter 2). Even then, research studies have shown that there are gender differences in the high-stakes examinations at the secondary level, as discussed in a later section of this chapter. If we are to think about inclusive practices in the classroom with regard to technology use, the questions to consider will be how much of students' confidence and attitude towards technology are influenced by their teachers' teaching approaches, and whether these translate to differences in mathematics achievement. The study described in this chapter takes a step in this direction and investigates students' preferences when they learn how to use the GC.

The structure of this chapter is as follows. First, the background context of Singaporean mathematics education is described. This is followed by a review of the literature on Singaporean calculator and gender studies, and how students learn to use advanced calculators (referring here to the GC and calculators with computer algebra system, CAS). The next section on methodology described the instruments used and method of data collection. Thereafter, the analysis and discussion section is presented, followed by the conclusion.

Background on Singaporean Mathematics Education Context

A comprehensive introduction to the policies and background context of the Singaporean education system and its mathematics curriculum is presented by Wong and Lee (2009). Singapore's general education system consists of 6 years of primary schooling and 4 or 5 years of secondary schooling. At the end of the secondary schooling, students take a national examination, of which a resultant aggregated score is calculated for entrance into postsecondary institutions. Postsecondary educational pathways include: 2 or 3 years of preuniversity schooling at a junior college or a centralised institute followed by university education, or tertiary education in a polytechnic or a vocational institution (see http://moe.gov.sg/education/landscape/).

At the end of the preuniversity schooling (grades 11 and 12), students sit for the Singapore-Cambridge General Certificate of Education (Advanced Level) (also called the GCE A-Level) examinations. The examination results are used for application into universities and tertiary education institutions. Detailed information about the high-stake examinations can be found in the Singapore Examination and Assessment Board website (http://www.seab.gov.sg). In general, students' performances in the various high-stake examinations determine the possible academic pathways they can take. The Singaporean government prides itself as having a meritocratic system in which entrance into secondary and postsecondary educational institutions are merit-based, although there are equity concerns raised by educators relating to social immobility in an environment with differentiated academic pathways and where parental income impact on the amount and kinds of resources and external support students have access to (e.g., see Forum letter replies, MOE, 23 February, 2011).

At the time of writing there were 13 Junior Colleges, which ran 2-year courses (equivalent to grades 11 and 12), one centralised institution, which ran a 3-year course and seven mixed level schools (grades 7 to 12), that all led to the GCE A-Level examination at the end of preuniversity education. There were schools which offered the International Baccalaureate and other equivalent qualifications; these were not the mainstream pathways and thus were not included in this study.

Under the 2006 revised A-Level curriculum, there were three levels of study for certain core subjects: Higher 1 (H1), Higher 2 (H2) and Higher 3 (H3). Access to and use of GCs (without CAS capacity) have been assumed in the H1, H2 and H3 GCE A-Level mathematics examinations since 2007. Students can choose to take either H1 or H2 mathematics, and high-ability students who took H2 mathematics could also take H3 mathematics, usually through invitations by their schools or recommendations by their teachers. H1 mathematics was a prerequisite for business, arts and accounting courses in Singaporean universities, and H2 Mathematics was a prerequisite for science and engineering courses (MOE 2007). The content, assessment and amount of curriculum time used at H1 level was half that of the H2 level. The use of GCs is expected in all the mathematics examinations, which comprised one (for H1) or two (for H2) 3-h written examinations at the end of the grade 12 academic year.

Singaporean junior colleges usually run a lecture-tutorial system rather than classroom teaching. The junior college students attend lectures for the various subjects in lecture theatres as a large group, as well as attend tutorials in classes of about 20–35. Since H1 and H2 mathematics are prerequisites for a number of faculties in Singaporean universities, almost all students take either H1 or H2 mathematics; high-ability students also take H3 mathematics. In addition, there are no curriculum-specific textbooks at the preuniversity level: students rely on lecture notes provided by teachers and reference books. Some calculator guidebooks may be prescribed as "textbooks" for reference.

In terms of the learning environment in Singaporean schools, Kaur (2004) provided an account of the mathematics curriculum framework, major educational initiatives and policies, as well as the schooling context and culture from the perspectives of heads of mathematics department in eight secondary schools. An emphasis on "mastery of knowledge and skills, critical and creative thinking, communication and problem solving" (Kaur 2004, p. 9) was described. Similar schooling culture and expectations can be assumed for preuniversities.

Calculator and Technology Use in Singaporean Mathematics Education

In terms of research on the use of technology in mathematics education, Ng and Leong (2009) provided a commentary and review of Singaporean studies. Most studies were on the affordances of technology in teaching and learning, as well as in relation to other factors in the instructional environment. The small scale and duration of these studies limited the generalisation of the findings (Ng and Leong 2009).

The introduction of GC in the mainstream mathematics A-level curriculum occurred during the review of the curriculum in 2006 and there was a scarcity of research about its use by Singaporean teachers. Before the implementation of GCs, Tan and Forgasz (2006) compared 33 Singaporean and 35 Victorian (Australian) senior secondary mathematics teachers' views about GCs, and found that Singaporean teachers were less certain about the usefulness of the GC and indicated less GC proficiency than Victorian teachers. It is speculated that the mandatory use of GCs in the Victorian high-stakes examinations played a part in the teachers' perceptions and use of the tool. In the same study, Tan (2005) also found that 90 % of the Singaporean teachers used computers in lectures or teacher demonstrations. Less than half of the teachers reported using computers in other teaching modes such as students working individually (45 %), students working in small groups (36 %), co-operative learning (23 %) and as a reward for students (13 %). The top three factors influencing the Singaporean teachers' computer use were found to be the suitability of the topic taught, adequate technical support and teachers' familiarity with the tool. It is suggested that the practices of Singaporean teachers reflected a didactical approach which focused on preparing students for the GCE A-Level examinations.

Ng (2006) conducted a study in 2003 on the use of the GC by junior college Further Mathematics (FM) students. FM was a higher-level mathematics subject aimed to prepare students for engineering and mathematics courses in university, and was removed when the new Advanced Level curriculum took effect in 2006. The GC was permitted in the FM grade 12 examinations in 2001 and the questions were said to be GC-neutral. Ng surveyed 190 students on three occasions, and the findings suggested that students using the GC performed better academically than those not using the GC. Ng (2006) contended that students' competency with the GC could be influenced by other factors such as access to the calculator, familiarity with its functions and the extent of exposure students have to its use, the latter being dependent on the teaching and learning environment.

In another study, a 6-month CAS intervention programme was conducted in a junior college in 2004 where two classes of high-achieving grade 11 students used

TI Voyage 200 calculators in their mathematics lessons (Ng et al. 2005). Students' attitudes towards CAS calculators (anxiety, confidence, liking and usefulness), measured by a 40-item questionnaire, were found to have improved significantly between pre- and posttreatment. The scores for the Usefulness and Liking subscales were highly correlated (Pearson's $r=0.826$), suggesting that the high-achieving Singaporean students were practical, liking the tool more the more it was perceived to be useful (Ng et al. 2005). Furthermore, the researchers found that students' initial enthusiasm with the CAS calculators diminished when they learnt that the calculators were not allowed in the examinations. In their written journals, students wrote that they were impressed by the symbolic, graphic and numeric capabilities of the CAS calculators and used them in a variety of ways, including checking answers, solving equations, proving identities, sketching graphs and simplifying mathematical expressions. They were dissatisfied with the limitations of the calculators, such as low screen resolution leading to distortion of graphs around asymptotes, inability to solve inequalities completely, calculator syntax different from conventional mathematics and limitations with solving equations involving complex roots. Despite this, students commented that the difficulties raised their awareness about writing mathematical expressions and about elements of graphing such as the domain of a function.

In summary, there is a scarcity of recent research on the GC in Singapore, particularly after its introduction into the mainstream senior secondary curriculum.

Gender Studies of Mathematics Education in Singapore

In terms of gender studies in Singapore, Kaur (1995) reviewed the literature and found that findings from studies generally concur with the larger literature in terms of males outperforming females in overall achievement of mathematics. She cited work by Leuar (1985) on 163 senior secondary students. It was found that although their grade 10 (national examination) achievements were similar, males outperformed females in the end of grade 11 mathematics test. Both males and females also believed that males would do better in mathematics. Other studies conducted at secondary levels also revealed that boys generally outperformed girls; and where otherwise, there were no gender differences. Kaur (1995) suggested that social factor might be implicated given that Singapore is a "traditionally male-dominated society in which mathematics and related skills are regarded as strictly male capabilities" (p. 133). More recently for PISA 2009, Singaporean males outperformed females by 5 points, with the gender difference being statistically significant (OECD 2010). This difference is smaller than the overall average of 12 points gender difference across the 65 participating countries and economies.

On the other hand, the result from earlier research studies and PISA 2009 that males outperformed females contradicts results from TIMSS. Data from TIMSS showed that the mean scores for grade 8 Singaporean girls were statistically significantly higher than Singaporean boys in the 2003, 2007 and 2011 studies

(Mullis et al. 2012). Soh and Quek (2001) analysed data from the TIMSS for four Asian nations, and found that on average Singaporean girls tended to outperform boys in both the seventh and eighth grades. They suggested that the reason why females did better than males in the TIMSS study was that there is a nonsexist learning environment in Singapore, "where teachers as well as parents, peers and society, expect boys and girls alike to do well in mathematics" (p. 337). In another study, Lim (2010) examined the performance of 984 students (500 males, 484 females) from a top Singaporean preuniversity in a three hour test (which was similar to the GCE A-level examination and required the use of GCs) and found that there was no significant difference in their achievement. However, from an attitude survey of these students, Lim found that males expressed more confidence in their mathematics ability than females and females had higher mathematics anxiety than males.

In addition, a survey of 1185 primary and secondary pre-service teachers' technological pedagogical content knowledge (TPACK) by Koh et al. (2010) revealed that generally male pre-service teachers scored significantly higher on technological knowledge, content knowledge and knowledge of teaching with technology, compared to female teachers. Although the group of pre-service teachers included nonmathematics teachers, the findings are consistent with that from another study reported by Ng (2003).

In Ng's (2003) study on secondary students' use of CAS reported earlier, he found that males scored higher than females in the mathematics tests using CAS. However, the difference was not statistically significant.

Overall, these findings demonstrated that other than the TIMSS findings, most other studies showed that Singaporean males scored higher than females in mathematics achievement and confidence, as well as technology confidence. Culturally there is a perception that mathematics and computers belonged to the male domain (see Chapter 6 of this volume by Forgasz and Leder). There was also a societal perception that mathematics is important subject necessary as a prerequisite for future education and all students, males and females, are expected to do well in mathematics (Soh and Quek 2001), at least at the secondary level. There were no studies found during the literature search that investigated gender difference when learning how to use technology for mathematics learning amongst Singaporean senior secondary students.

How Students Learn How to Use Technology

Studies have shown that familiarity with the technology is important to harness its potential for learning mathematics. Learning how to use the technology can be a challenge for some students and findings from some studies (e.g. Clarke et al. 2005) suggest that students' mathematics learning with technology may be more effective if they had prior training on the skills associated with the technology used. There are some evidence which suggest that the handheld calculator can be underused, especially "when students are not sure how to use the technology as a tool in their work or when they are unsure how much written work is required" (Burrill et al. 2002, p. ii). With advanced calculators like GC or CAS calculators, students may employ more superficial pragmatic strategies such as trying out a large number of guesses more efficiently and finding the intercept of a graph through repeated pressing of the cursor key to trace the coordinates along the graph (Ruthven 1996). Other researchers have cautioned about students' over-reliance on technology, and over-use of the calculator to the point that it is used "with little critical analysis of the [calculator] results" (Burrill et al. 2002, p. v).

However, there is not much information about how students *learn to use the technology*, in other words, the acquisition of technological skills or insight. For example, how do students learn how to use the GC and develop the GC skills? In studies on students' GC use, students' familiarity with the tool was associated with the time spent on learning with the tool (e.g. Burrill et al. 2002). Rather than looking at the intended or the attained curriculum, we need to examine more closely the implemented curriculum, in terms of the instructional practices of how students learn how to use the calculator in order to become familiar with it. There might also be gendered responses to these practices (e.g. see Boaler 2007) that bears investigating.

The issue of how students learn to use the calculator is complex and difficult to isolate because there is a dialectic relationship between mathematical and technical demands in calculator use. Technical demands increase with the introduction of new mathematical concepts which require the learning of new commands. At the same time, a certain level of mathematical understanding is required in order to use the calculator effectively. This inter-relatedness between mathematical and technical knowledge has also been pointed out by other researchers. In characterising the nature of obstacles faced by preuniversity students using CAS calculators, Drijvers (2000) proposed that there was "a technological, machine-related component, but dealing appropriately with them also requires mathematical insight" (p. 205). Pierce and Stacey (2004) described the importance of having algebraic insight (having knowledge of algebra and ability to link representations) in deciding when to use CAS technologies and in entering expressions into CAS. Burrill et al. (2002) believed that "in order to use the calculator successfully, students need to be familiar with the mathematics surrounding the task at hand and recognize how the limitations of the calculator can inhibit understanding of the mathematics" (p. 20). Dahland and Lingefjärd (1996) also espoused the importance of students having the "technical insight to be able to interpret the information given on a graphics screen, and… [they] must also have a sufficiently good mathematical understanding to realize the connection between the current problem and the possibilities given by the tool" (p. 31). Thus, it is not surprising that high achievement in mathematics was found to be associated with high levels of mathematics and technology confidence and a strongly positive attitude towards learning mathematics with technology, whereas low achievement in mathematics was found to be associated with low levels of mathematics and technology confidence and a negative attitude towards learning mathematics with technology (Barkatsas et al. 2009).

Familiarity with technological tools also relate to gender differences, since it is associated with having technological confidence and competence. From research, males tended to be more confident in and have greater confidence in technology than females (e.g. Barkatsas et al. 2009; Tan and Forgasz 2011). This might not be considered surprising because of two main social (and perhaps academic) perceptions: (1) GC is a technology, and boys tended to be better at technology (see Forgasz & Leder, Chapter 6); and (2) GC is a visual and kinaesthetic tool and from past research, and there may be gender differences associated with visual and kinaesthetic skills (see Leder 1990).

Overall, prior gender studies in the Singaporean context generally focus on students' achievement and attitudes (e.g. Ng 2003), rather than the other student factors such as their instructional preferences when learning how to use the GC. Hence, in the study reported here, Singaporean senior secondary students' learning preferences for using GC are investigated and examined for any gender differences. The research questions are:

- What are students' confidence and competence with the GC?
- What instructional methods are (perceived to be) useful to students when they learn how to use the GC? What are students' most-preferred methods?
- What instructional methods are (perceived to be) used by their teachers?
- Are there gender differences in the above?

The above sections presented a review of the literature on Singaporean calculator and gender studies in mathematics education, as well as a discussion of how students learn to use technology. In the next section, the methods and analyses used in this study are presented.

Methodology

Description of the Instrument

The study, part of a larger study, was conducted in Singapore using an online survey created using the SurveyMonkey platform. Invitations were sent to students through schools to participate in the study.

Measuring Students' Mathematics and GC Competency Self-Ratings

In Singapore, there are no common grade 11 and 12 assessments across schools other than the A-level examinations at the end of grade 12. Since actual measures of students' mathematical competence could not be obtained, and there are no measures of GC competencies, items tapping students' self-perceptions of their mathematics and GC competencies were used. Past research studies have found that there was a high correlation between students' self-reports of competency and their actual achievement (Hattie 2009). In this study, the two items on mathematics and GC competency self-ratings were: "Currently for mathematics I consider myself ___" and "In terms of GC skills, I consider myself ___". A Likert-type response format was used and scored in ascending order: 1 = "Weak", 2 = "Below Average", 3 = "Average", 4 = "Above Average", 5 = "Excellent".

Measuring Students' Confidence of GC

Students were asked to indicate their agreement with two statements: "I enjoy using GC to learn maths." and "I feel confident doing maths with GC." using 5-point Likert response format (1 = "Strongly Disagree", 2 = "Disagree", 3 = "Neutral", 4 = "Agree", 5 = "Strongly Agree").

Measuring Students' Most-preferred Method of Learning How to Use the GC

Students were asked to select, out of ten methods, their most-preferred method when learning how to use the GC to solve mathematics problems. The ten methods were derived based on the author's own teaching experience in Singapore, as well as readings of learning styles and modality preferences (e.g. the visual, aural, read-write, and kinaesthetic modes by Fleming 2006). Besides the methods listed, there was also an "other" option where students could enter their most-preferred method if it was not found on the list:

- See my teacher's demonstration in class
- See the steps my friends show me on their GC
- Look at the GC screen captures in notes, textbooks or manual
- Discuss answers with my friends
- Listen to a teacher who explains the steps and concepts clearly and thoroughly
- Listen to a teacher who reads out the steps given in notes, textbooks or manual
- Copy down the steps my teacher writes on the board
- Make my own notes
- Try out the steps on the GC at the same time I see a demonstration or hear an explanation or read the instructions
- Try the buttons out and play around with the GC

Measuring Perceived Effectiveness of the Methods of Learning How to Use the GC and Perceived Instructional Use of These Methods

For each of the 10 methods listed above, students were asked whether the method was helpful or not to their learning of the GC ("Yes it definitely helps" or "No it doesn't really help"), and to select those methods which their teacher used. The items are reproduced below:

For each of the following, indicate if it helps you learn how to use a graphing calculator to solve maths problems.	Yes, it definitely helps	No, it doesn't really help
See my teacher's demonstration in class	☐	☐
See the steps my friends show me on their calculators	☐	☐
Look at the calculator screen captures in notes, textbooks or manual	☐	☐
Discuss answers with my friends	☐	☐
Listen to a teacher who explains the steps and concepts clearly and thoroughly	☐	☐
Listen to a teacher who reads out the steps given in notes, textbooks or manual	☐	☐
Copy down the steps my teacher writes on the board	☐	☐
Make my own notes	☐	☐
Try out the steps on the calculator at the same time I see a demonstration or hear an explanation or read the instructions	☐	☐
Try the buttons out and play around with the calculator	☐	☐

Which one or more of the following ways has your teacher used when teaching how to use the graphing? [Please tick where appropriate]

☐ (a) Provide a demonstration
☐ (b) Let students demonstrate to the whole class
☐ (c) Refer to the calculator screen captures shown in notes or textbooks or manual
☐ (d) Let students discuss answers with one another
☐ (e) Explain the steps and concepts clearly and thoroughly
☐ (f) Read out the steps given in notes, textbooks or manual
☐ (g) Write out the steps on the board
☐ (h) Ask you to make your own notes
☐ (i) During a demonstration ask you to follow the steps as shown
☐ (j) Encourage you to play around with the calculator
☐ OTHERS (Please describe what activities your mathematics teacher does in your class)

As can be seen, not all "teaching" methods match exactly with the "learning" methods. For example, for "learning" method (b) "see the steps my friends show me on their calculators" students may discuss and show one another the steps on their calculators, which may or may not be facilitated by the teacher. The "teaching" method (b) "let students demonstrate to the whole class" was only one teaching strategy in which a student might see another student's GC steps. Note that peer discussion was covered in "teaching" method (d). Nonetheless, most of the methods from the two questions were meant to match to a certain degree.

Sample

Four schools agreed to participate in the study and sent an email to their students. In total, there were 964 Singaporean students ($N_F = 606$ or 62.9 % females, $N_M = 358$ or 37.1 % males) from the four Singaporean preuniversities who responded. There were slightly more grade 11 ($N = 517$, 53.6 %) than grade 12 ($N = 409$, 42.4 %) students, and a small number of grade 10 ($N = 38$, 3.9 %) students from schools with integrated programme (grade 7–12 or grade 9–12 programmes). Most of the students took the H2 mathematics subject ($N = 932$, 96.7 %) and a small number took the H1 mathematics subject ($N = 16$, 1.7 %). Almost all students used Texas Instrument calculators: TI84+ ($N = 905$, 93.9 %) or TI 83+ ($N = 53$, 5.5 %). Nine students (0.9 %) indicated that they used Casio FX9860G calculator. The numbers add up to more than 100 % since a few students owned more than one GC.

Data Analyses

Data were analysed using Statistical Package for Social Sciences (SPSS v. 20) software. Frequency distributions, *t*-tests and independent groups Chi-square tests were used to compare gender differences in the variables measured by the survey.

Analysis and Discussion

Students' Mathematics and GC Competencies, and GC Confidence and Gender Differences

Using *t*-tests, statistically significant gender differences were found for students' mathematics competency self-rating (MSR), calculator competency self-rating (CalSR) and calculator confidence (Cal_Conf). Table 5.1 shows the differences in the mean scores between male and female participants. It can be seen in Table 5.1 that males scored significantly higher than females in mathematics competency, calculator competency and calculator confidence. Female students generally rated themselves below average in mathematics and calculator competencies, and males rated themselves slightly above average. Both agreed that they were confident in using the GC, but males agreed more strongly than females that they were confident, with an effect size of 0.18. That females tended to be less confident is consistent with the other research studies (e.g. see Chapter 2 by Leder and Lubinski), although the effect sizes were considered small.

Students' Most-Preferred Methods of Learning How to Use the GC, and Gender Differences

Table 5.2 shows the frequency and percentages of students who selected their methods which they most prefer when learning how to use the GC. It can be seen in Table 5.2

Table 5.1 Gender differences in students' MSR, CalSR and Cal_Conf

	Gender	*N*	Mean	*t*	Effect size *r*[a]
MSR (M>F)	Female	605	2.79	−4.01*	0.13
	Male	358	3.08		
CalSR (M>F)	Female	604	2.87	−3.59*	0.12
	Male	358	3.07		
Cal_Conf (M>F)	Female	588	3.08	−4.64*	0.18
	Male	349	3.40		

MSR mathematics competency self-rating, *CalSR* calculator competency self-rating, *Cal_Conf* calculator confidence, *M* male, *F* female

$*p<0.001$

[a] Effect size for *t*-test $r = \sqrt{\dfrac{t^2}{t^2 + df}}$ (Field 2005)

that certain methods were more frequently cited than others. For both males and females, methods (a), (e), (i) and (j) were preferred by about 10 % or more of the students. There were gender differences seen for methods (a), (i) and (j), with a difference of 5 % or higher between the percentages of males and females who most preferred the methods. Amongst all the methods, the highest percentages of females ($F=46.6\,\%$, $M=35.0\,\%$) most preferred to learn to use the GC by method (i) "trying out the steps on the GC the same time they see a demonstration, listen or read the instructions", and higher percentage of females than males preferring this method. On the other hand, higher percentages of males than females most prefer method (a) ($F=17.9\,\%$, $M=22.9\,\%$; "see the teacher's demonstration") and method (j) ($F=10.6\,\%$, $M=17.5\,\%$; "try the buttons out and play around with the GC"). Method (e) ("listen to a teacher who explains the steps and concepts clearly and thoroughly") was preferred by roughly similar percentages of females (9.7 %) and males (10.0 %).

If we look at Table 5.2, we can see that firstly there are some methods that were preferred by higher percentages of students, compared to other methods, in particular method (i) was the most popular amongst both males and females; secondly for each of the method, there are generally similar percentages of males and females who most preferred that method; and thirdly there are a few methods with differences of more than 5 % between the percentages for males and females, namely methods (a), (e), (i) and (j). The results show that there are gender differences in students' most-preferred methods of learning how to use the GC to solve mathematics problems, with a higher percentage of females than males most preferring method (i) (a difference of 11 %) and higher percentages of males than females most preferring methods (a) (a difference of 5 %) and (j) (a difference of 7 %). Based on the results, a possible conclusion would be that the Singaporean students tended to employ visual and kinaesthetic modes when learning how to use the GC, with method (i) of "trying out the steps at the same time they watch a demonstration, listen to an explanation or read the instructions"—a form of guided practice—being the most popular. Male students were more likely than females to learn how to use the GC through watching a demonstration (method (a)) and through playing around with the GC (method (j)), whilst females were more likely than males to learn how to use the GC through trying out the steps with guided instruction (method (i)).

Table 5.2 Frequencies and percentages of Singaporean females and males who selected (a)–(j) as their most-preferred method of learning how to use the GC

	Female			Male			Difference in %
	Count	Expected count	% within gender	Count	Expected count	% within gender	(*F–M*%)
(a) See my teacher's demonstration in class	107	118.0	17.9	80	69.0	22.9	−5.0
(b) See the steps my friends show me on their calculator	23	22.1	3.9	12	12.9	3.4	0.5
(c) Look at the calculator screen captures in notes, textbooks or manual	34	34.7	5.7	21	20.3	6.0	−0.3
(d) Discuss answers with my friends	5	6.9	0.8	6	4.1	1.7	−0.5
(e) Listen to a teacher who explains the steps and concepts clearly and thoroughly	58	58.7	9.7	35	34.3	10.0	−0.3
(f) Listen to a teacher who reads out the steps given in notes, textbooks or manual	1	1.3	0.2	1	0.7	0.3	−0.1
(g) Copy down the steps my teacher writes on the board	14	11.4	2.3	4	6.6	1.1	1.2
(h) Make my own notes	14	13.3	2.3	7	7.7	2.0	0.3
(i) Try out the steps on the calculator at the same time I see a demonstration or hear an explanation or read the instructions	278	252.4	46.6	122	147.6	35.0	11.0
(j) Try the buttons out and play around with the calculator	63	78.3	10.6	61	45.7	17.5	−6.9
Total	597	597	100	349	349	100	

Students' Perceived Usefulness of the Methods of Learning How to Use the GC

Students were given the set of ten methods and they indicated whether they found each method useful or not. They were also asked to select the methods which their teacher used. The methods listed as what their teacher used were modified from the list of methods students employed to learn how to use the GC.

The responses to these questions revealed more detail than the question described previously where students selected the most-preferred method out of the set of ten methods. Comparing their answers to the questions, there are four cases for each method:

- Students said a "learning" method is definitely helpful AND their teacher used the corresponding "teaching" method.
- Students said a method is definitely helpful AND did not select that method when asked to select the methods their teacher used (I assumed that it was perceived by students that their teacher did not use it).
- Students said a method was not really helpful AND their teacher used it.
- Students said a method was not really helpful AND their teacher did not use it.

Of the four cases, the first two would be the most interesting to explore, in the sense that we want to look at how inclusive the teachers' instructions were, based on students' perceptions. A high percentage of case (1) is desirable. Of course, for a particular "learning" method, teachers could create various opportunities for students to employ the method, rather than use the particular "teaching" method per se.

Table 5.3 shows the percentages of males and females that fell into cases (1) and (2), by methods. Chi-square tests of independence were performed on each method to compare the distribution of females and males across the four cases. The sum of the percentages for cases (1) and (2) is the percentage within gender of students who said that the method definitely helped them learn how to use the GC. As can be seen in Table 5.3, the tests revealed that there are statistically significant gender differences for methods (b), (e), (g), (h), and (i). Note that percentages for cases (3) and (4) are not included.

For method (b), roughly equal percentages of females and males ($F=5.9+83.1=89.0\,\%$, $M=9.3+81.1=90.4\,\%$) responded that seeing the steps on their friends' calculators was helpful to them, but slightly higher percentages of females than males ($F=83.1\,\%$, $M=81.1\,\%$) did not perceive that their teachers let students demonstrate to the class. There are other opportunities for students to show one other their calculator steps (e.g. getting students to discuss answers with one another, method (d)), however, there was also a low percentage of students who said that their teachers used method (d) (adding percentages for cases (1) and (3): $F=26.6\,\%$, $M=29.8\,\%$, not shown in Table 5.3).

For method (e), again there were roughly equal percentages of females and males ($F=87.2\%$, $M=89.9\%$) responded that listening to the teacher who explained the steps and concepts clearly and thoroughly helped them learn how to use the GC. However, higher percentages of males than females ($F=28.5\%$, $M=33.8\%$) perceived that their teachers did not explain the steps and concepts clearly and thoroughly.

For method (g), 60.2 % of females said that copying the steps the teacher wrote on the board was useful, but only 50.4 % of the males said the same. There was also a much higher percentage of females than males who found the method useful but that their teacher did not write the steps on the board ($F=32.1\,\%$, $M=24.9\,\%$).

For method (h), 68.7 % of females said that making their own notes helped them learn how to use the GC to solve mathematics problems, but only 60.3 % of males

Table 5.3 Percentages of Singaporean females and males for cases (1) and (2) for each method, and gender differences

Method and cases	Female		Male		Chi-square test
	Case 1[a] (%)	Case 2[b] (%)	Case 1 (%)	Case 2 (%)	
Method (a) Learning: see my teacher's demonstration in class; Teaching: provide a demonstration	87.9	6.8	89.0	6.8	NS
Method (b) Learning: see the steps my friends show me on their calculators; Teaching: let students demonstrate to the whole class	5.9	83.1	9.3	81.1	$\chi^2(3,N=953)=9.77$, $p=0.033$
Method (c) Learning: look at the calculator screen captures in notes, textbooks or manual; Teaching: refer to the calculator screen captures shown in notes or textbooks or manual	64.4	13.7	61.4	14.4	NS
Method (d) Learning: discuss answers with my friends; Teaching: let students discuss answers with one another	52.8	23.2	48.6	24.1	NS
Method (e) Learning: listen to a teacher who explains the steps and concepts clearly and thoroughly; Teaching: explain the steps and concepts clearly and thoroughly	58.7	28.5	56.1	33.8	χ^2 (3,N=958)=9.38, $p=0.025$
Method (f) Learning: listen to a teacher who reads out the steps given in notes, textbooks or manual; Teaching: read out the steps given in notes, textbooks or manual	23.7	22.0	25.5	24.4	NS
Method (g) Learning: copy down the steps my teacher writes on the board; Teaching: write out the steps on the board	28.1	32.1	25.5	24.9	χ^2 (3, N=940)=15.0, $p=0.02$
Method (h) Learning: make my own notes; Teaching: ask you to make your own notes	7.6	61.1	7.7	52.6	χ^2 (3, N=941)=11.6, $p=0.009$
Method (i) Learning: try out the steps on the calculator at the same time I see a demonstration, hear an explanation or read the instructions; Teaching: during a demonstration ask you to follow the steps as shown	72.8	22.2	64.2	29.9	χ^2 (3, N= 958)=7.94, $p=0.047$

Table 5.3 (continued)

Method and cases	Female		Male		Chi-square test
	Case 1[a] (%)	Case 2[b] (%)	Case 1 (%)	Case 2 (%)	
Method (j) Learning: try the buttons out and play around with the calculator; Teaching: encourage you to play around with the calculator	42.0	33.5	45.1	30.9	NS

[a] Case 1: method useful, teacher used it
[b] Case 2: method useful, teacher did not use it

said the same. There was also much higher percentage of females than males who found the method useful but that their teacher did not ask them to make their own notes (F=61.1 %, M=52.6 %).

For method (i), roughly the same percentage of females and males (F=95.0 %, M=94.1 %) said that trying out the steps on the GC at the same time as they watched a demonstration was helpful. However, much higher percentage of females than males (F=72.8 %, M=64.2 %) said their teacher encouraged them to try out the steps during demonstrations, whereas higher percentage of males than females (F=22.5 %, M=29.9 %) perceived that their teacher did not encourage them to try out the steps during demonstrations.

Overall, there are differences between what students' found useful, and what methods they said their teachers used. Certain methods which students indicated as helpful to their GC learning, such as method (b) (letting students demonstrate to the class), method (d) (letting students discuss answers with one another), method (h) (asking students to make their own notes) and method (j) (encouraging students to play around with the calculator), were not well-employed by teachers. As these methods are student-centred approaches, the findings indicate that Singaporean teachers need to consider employing more student-centred approaches. Note that although the methods ((b), (d), (h) and (j)) were found to be useful by a majority of students, they were cited by only small percentages of students as their most-preferred method of learning how to use the GC (see Table 5.2). It could be that these methods were not well supported by the teachers, or that students simply preferred other methods over these.

It is interesting to note that there were gender differences only in the percentages of students who found methods (g) and (h) useful. For all the other methods, there were similar percentages for females and males who found the methods useful. Both methods (g) and (h) were of a read-write modality: copying the steps teachers wrote on the board and making their own notes. For each method, about 10 % more females than males found the method useful and there are similar percentages of females and males that perceived their teachers to be using the methods. It would seem that the "extra" 10 % of females who found the method to be useful indicated that their teachers did not use the methods. A conclusion drawn from this is that there was around a tenth more of the females than males who found methods (g) and (h) useful, but perceived that their teachers did not employ these methods. Hence, it seemed that more females' than males' needs (to write down the steps to solve mathematics problems and make their own notes) were not perceived to be met by their teachers.

It is interesting that for method (i) much higher percentages of females than males said that their teacher encouraged them to try out the steps as they watched a demonstration. It could be that teachers treated females and males differently and encouraged more females to try out the steps as they watched a demonstration (gendered practices by teachers) or that females were more likely than males to perceive their teachers to be encouraging (gendered perceptions by students). The second reason may be less likely than the first since for method (j), slightly lower (rather than higher) percentages of females than males perceived the teacher as encouraging them to try out the steps and play around with the GC.

From the responses for methods (e) and (i), it may be concluded that more males' than females' needs for these two methods were not met. There were similar percentages of males and females who said that the methods were helpful, but significantly higher percentages (5–8 %) of males than females said their teachers did not employ these methods. It would be interesting to find out if which group of students belonged in the same class or taught by the same teacher; unfortunately class and teacher information was not captured in the anonymous survey.

Implications and Conclusions

Overall, the findings raised a few questions to consider.

Are Students' Learning Needs with Regard to GCs not Met?

There were certain methods (from Table 5.3) which students indicated as helpful to their learning how to use the GC but did not indicate that their teachers used the corresponding methods. These methods—(b) seeing the GC steps on their friends' calculators, (d) discussing answers with their friends, (h) making their own notes and (j) playing around with the calculators—seemed to be more student-centred. In an examination-oriented environment where high-stakes examinations' results determine university entrance, the focus is on individual achievement. This could lead to teacher-centred approaches being favoured and less focus being placed on cooperative learning strategies by teachers.

Are There Higher Proportions of Female Students' Needs not Met, Compared to Males?

There are two methods that were helpful to higher proportions of females than males, and the methods were also perceived by more females than males to be not employed by teachers. These two methods are: (g) copying down the steps which the teacher wrote on the board and (h) writing their own notes. Generally, in Singaporean junior colleges the teachers prepare notes and handouts for students; there is no curriculum-specific textbook used. Perhaps writing their own notes and taking down notes as the teacher explained the steps helped some females more than males because females were more likely than males to seek understanding and personal connection with the concepts they were learning (Boaler 2007). There might also be gendered teaching practices in that teachers might be more likely to encourage more females than males to try out the steps as they watched a demonstration (method (i)), and more likely to encourage more males than females to try and play with the GC in order to learn its functions (method (j)). Further studies can be conducted to investigate whether teachers have gendered practices, and if any, how they impacted males and females.

What Kinds of Questions Should We Ask About Gender and Technology that Help Us to be More Inclusive?

There are two sets of questions investigated in the study presented here. One set of questions is focused on learning outcomes by gender, such as students' competency and confidence in mathematics and technology, as well as their most-preferred learning method. The other set of questions is focused on the learning process and environment, by looking at what students said about the usefulness of various learning methods, and what methods they said their teachers used. If we answer the first set of questions, we might risk perpetuating the essentialism view of gender (see Boaler 2007 for a discussion). For example from Table 5.2, we can see that a higher percentage of females than males most-preferred method (i) and therefore conclude that females were more likely than males to prefer a guided practice approach. Males were also more likely to most-preferred method (j): to play around with the GC, a finding which perpetuates the view that males were more kinaesthetic and had higher confidence in using technology. In terms of inclusivity, method (i)—the method cited by the most number of students—was already employed by most teachers (71 % of the 964 students responded said their teachers asked them to follow the GC steps during a demonstration). This seemed to indicate that the teachers' practices generally did cater to students and in particular to female students, since a higher percentage of females than males most-preferred method (i). However, if we answer the second set of questions, we see a slightly different story: that there may be certain student-centred approaches that were not well employed by teachers; that there were some methods which higher proportions of females' than males' needs were not met. It might be further speculated that this might explain why there were higher percentages of females than males who relied on method (i) as their most-preferred method of learning how to use the GC. Females whose preferred method(s) were not supported by their teachers may adopt other well-supported methods as their *most-preferred* method. This might be similar to the example Boaler (2007) gave of Dweck's labelling of high-achieving girls' tendency to avoid high-risk situation as "maladaptiveness". Boaler argued that the "bright girls" were actually highly adaptive to their environment, which was one with high-stakes competition and where correct answers were valued. Maladaptiveness became the label with which characterised some girls, rather than describing their behaviour as a coproduct of the schooling and classroom environment in which they were situated. It may be argued that the second set of questions investigated in this study revealed more about gender as a situated response to the learning environment and context rather than gender as a set of characteristics and attributes, and thus is a more inclusive way of investigating gender (Boaler 2007).

In the national high-stake examinations, Singaporean males tended to do better than females (Kaur 1995). Soh and Quek (2001) suggested that the sociocultural environment and practices influenced male and female students differently. Dindyal (2008) highlighted the paradox of the Asian learner (Biggs 1994) as one of the possible factors influencing gender differences. The paradox was that students from Confucian-heritage cultures such as China, Hong Kong, Taiwan, Singapore, Korea and Japan, are perceived as passive rote learners and taught in classroom conditions not viewed as conducive to good learning (e.g., large class size, expository methods, and high prominence of assessment). Yet these students showed high level of understanding and outstanding performance in international comparative studies. Biggs (1994) argued that the paradox is due to different cultural perspectives and interpretations of education and learning. Hence, culture also plays an important part in understanding the context in order to make further conjectures about the findings of this study. The Singaporean education system, being meritocratic, is highly competitive and there is a strong emphasis placed on individual effort and examination performance. Despite the call by policy makers to emphasise student-centred quality teaching and learning, it would be a challenge to change the mindset and practices of the teachers.

> Why should a teacher use two hours to allow students discover a concept for themselves when he can use one hour to teach it and another hour to drill the students to practice-perfection, especially when it is likely that the examinations will test the latter than the formal? (Ng 2008, p. 12)

In an environment where there are tensions between the strong emphasis on attaining excellent educational achievement and the call for a creative student-centred pedagogy, Singaporean primary and secondary pre-service teachers' were found to be generally inclined to agree with constructivist beliefs, but their constructivist beliefs were strongly correlated with both constructivist and traditional use of information and communication technology (ICT) in schools (Chai 2010). Given the context where teachers have to prepare students for high-stakes examinations in a limited time, they are more likely to fall back to traditional teacher-centred practices despite having constructivist beliefs (see, e.g., Lim and Chai 2008). Consequently, in such a competitive environment females might be seen to be at a greater disadvantage than males (e.g., Boaler 2002). On the other hand, females might also be seen as benefitting from the didactical structured method of teaching (e.g. rule-following and rote learning approach, Fennema and Peterson 1985). With the use of technology such as the GCs in high-stake examinations, mathematics teachers planning effective and inclusive practices need to take into account a host of factors including students' confidence and competence with the technology, attitude and competency in mathematics and their learning preferences, as well as the students' responses towards different teaching approaches. In short, it would seem that a student-centred pedagogy that promotes quality learning of every individual, rather than teaching to different "labels" such as gender (for other discussions of labels, see Chapter 12 of this volume by Bishop and Kalogeropoulos) of the students, might be more effective.

Girls have been found to be less confident and competent than boys in their use of technology for learning mathematics. The conventional discourse is an essentialist one (Boaler 2007). In investigating students' learning preferences, it is hoped that future research on gender and technology in mathematics education can be reconsidered in the same vein that is argued by Boaler—that is to consider the way the learning environment and the sociocultural context in which students learn influence the way they use technology to learn mathematics, and that gender differences may arise out of their responses to their learning environments. We should consider the constraints and affordances that the learning environment and the sociocultural context provide rather than attributing the differences in outcome variables to purely gender characteristics. A final note is that, although investigating students' responses to their learning environment may be seen to be more conducive to qualitative studies, there is a place for the quantitative studies such as the one presented in this and other chapters (e.g., Chapter 2 by Leder and Lubienski) to complement qualitative studies and contribute to the knowledge in the field.

References

Australian Association of Mathematics Teachers (AAMT). (1996). Statement on the use of calculators and computers for mathematics in Australian schools. http://www.aamt.edu.au/Publications-and-statements/Position-statements/Calculators-and-Computers/%28language%29/eng-AU. Accessed 24 June 2014.

Barkatsas, A., Kasimatis, K., & Gialamas, V. (2009). Learning secondary mathematics with technology: Exploring the complex interrelationship between students' attitudes, engagement, gender and achievement. *Computers and Education, 52*(3), 562–570.

Biggs, J. B. (1994). What are effective schools? Lessons from East and West. *The Australian Educational Researcher, 21,* 19–40.

Boaler, J. (2002). *Experiencing school mathematics: Traditional and reform approaches to teaching and their impact on student learning*. Mahwah: Lawrence Erlbaum.

Boaler, J. (2007). Paying the price for "sugar and spice": Shifting the analytical lens in equity research. In N. S. Nasir & P. Cobb (Eds.), *Improving access to mathematics: Diversity and equity in the classroom* (pp. 24–36). New York: Teachers College Press.

Burrill, G., Allison, J., Breaux, G., Kastberg, S., Leatham, K., & Sanchez, W. (2002). *Handheld graphing technology in secondary mathematics: Research findings and implications for classroom practice*. Dallas: Texas Instruments.
Chai, C. S. (2010). The relationships among Singaporean pre-service teachers' ICT competencies, pedagogical beliefs and their beliefs on the espoused use of ICT. *The Asia-Pacific Education Researcher, 19*(3), 387–400.
Clarke, T., Ayres, P., & Sweller, J. (2005). The impact of sequencing and prior knowledge on learning mathematics through spreadsheet applications. *Educational Technology Research and Development, 53*(3), 15–24.
Dahland, G., & Lingefjärd, T. (1996). Graphing calculators and students' interpretations of results: A study in four upper secondary classes in Sweden. *Nordisk Matematikkdidaktikk, 4*(2/3), 31–50.
Dindyal, J. (2008). An overview of the gender factor in mathematics in TIMSS-2003 for the Asia-Pacific region. *ZDM, 40,* 993–1005. doi:10.1007/s11858-008-0111-2.
Drijvers, P. (2000). Students encountering obstacles using a CAS. *International Journal of Computers for Mathematical Learning, 5,* 189–209.
Fennema, E., & Peterson, P. (1985). Autonomous learning behavior: A possible explanation of gender-related differences in mathematics. In L. C. Wilkinson & C. B. Marrett (Eds.), *Gender influences in classroom interactions* (pp. 17–35). New York: Academic Press.
Field, A. (2005). *Discovering statistics using SPSS* (2nd ed.). London: SAGE.
Fleming, N. D. (2006). *Teaching and Learning style: VARK strategies*. Christchurch: N. D. Fleming.
Forgasz, H., & Tan, H. (2010). Does CAS use disadvantage girls in VCE mathematics? *Australian Senior Mathematics Education Journal, 24*(1), 25–36.
Forgasz, H., Vale, C., & Ursini, S. (2010). Technology for mathematics education: Equity, access and agency. In C. Hoyles & J.-B. Lagrange (Eds.), *Mathematics education and technology—rethinking the terrain* (pp. 251–284). New York: Springer.
Gutiérrez, R. (2007). (Re)defining equity: The importance of a critical perspective. In N. S. Nasir & P. Cobb (Eds.), *Improving access to mathematics: Diversity and equity in the classroom* (pp. 37–50). New York: Teachers College Press.
Hattie, J. C. (2009). *Visible learning: A synthesis of over 800 meta-analyses relating to achievement*. London: Routledge.
Kaur, B. (1995). Gender and mathematics: The Singapore perspective. In P. Rogers & G. Kaiser (Eds.), *Equity in mathematics education: Influences of feminism and culture* (pp. 129–134). London: Falmer.
Kaur, B. (2004). *Teaching of mathematics in Singapore schools*. Proceedings of the 10th International Congress on Mathematical Education, Denmark. (http://www.icme10.dk/proceedings/pages/regular_pdf/RL_Berinderjeet_Kaur.pdf).
Koh, J., Chai, C., & Tsai, C. (2010). Examining the technological pedagogical content knowledge of Singapore pre-service teachers with a large-scale survey. *Journal of Computer Assisted Learning*, 1–11. doi:10.1111/j.1365-2729.2010.00372.x.
Leder, G. (1990). Gender differences in mathematics: An overview. In E. Fennema & G. C. Leder (Eds.), *Mathematics and gender* (pp. 10–26). New York: Teachers College, Columbia University.
Leuar, B. C. (1985) *Sex differences and mathematical ability*. Unpublished assignment during the period of induction, Singapore, Institute of Education.
Lim, S. Y. (2010). Mathematics attitudes and achievement of junior college students in Singapore. In L. Sparrow, B. Kissane, & C. Hurst (Eds.), *Shaping the future of mathematics education: Proceedings of the 33rd annual conference of the mathematics education research group of Australasia* (pp. 681–689). Fremantle: MERGA.
Lim, C. P., & Chai, C. S. (2008). Teachers' pedagogical beliefs and their planning and conduct of computer-mediated classroom lessons. *British Journal of Educational Technology, 39*(5), 807–828.

Ministry of Education Singapore (MOE). (27 Aug 2007). Breadth and flexibility: The new "A" level curriculum 2006. http://www.moe.gov.sg/cpdd/alevel2006/index.htm. Accessed 24 June 2014.

Ministry of Education Singapore (MOE). (23 Feb 2011). Forum letter replies: Singapore's meritocratic education system promotes social mobility. http://www.moe.gov.sg/media/forum/2011/02/singapores-meritocratic-education-system-promotes-social-mobility.php. Accessed 24 June 2014.

Mullis, I. V. S., Martin, M. O., Foy, P., & Arora, A. (2012). *TIMSS 2011 international results in mathematics*. USA: TIMSS & PIRLS International Study Center, The Netherlands: IEA.

Ng, E. H. (2008). Launch of MP3. http://www.moe.gov.sg/media/speeches/2008/08/05/opening-address-by-dr-ng-eng-h-1.php. Accessed 24 June 2014.

Ng, W. L. (2003). Effects of computer algebra system on secondary students' achievement in mathematics: A pilot study in Singapore. *International Journal of Computer Algebra in Mathematics Education, 10*(4), 233–248.

Ng, W. L. (Dec 2006). Use of graphing calculators in pre-university further mathematics curriculum. In L. H. Son, N. Sinclair, J. B. Lagrange, & C. Holyes (Eds.), *Proceedings of the ICMI 17 study conference: Background papers for the ICMI 17 Study* (Vol. 1, pp. 120–130). (http://www.math.msu.edu/~nathsinc/ICMI/papers/GroupeC.pdf).

Ng, W. L., & Leong, Y. H. (2009). Use of ICT in mathematics education in Singapore: Review of research. In K. Y. Wong, P. Y. Lee, B. Kaur, P. Y. Foong, & S. F. Ng (Eds.), *Mathematics education: The Singapore journey* (pp. 301–318). Singapore: World Scientific.

Ng, W. L., Kwee, T. C., Lau, H. S., Koh, Y.-H., & Yap, Y. S. (2005). Effects of using a computer algebra system (CAS) on junior college students' attitudes towards CAS and achievement in mathematics. *The International Journal for Technology in Mathematics Education, 12*(2), 59–72.

OECD. (2010). PISA 2009 results, volume I, what students know and can do: Student performance in reading, mathematics and science, Figure I.3.12. http://dx.doi.org/10.1787/888932343152. Accessed 24 June 2014.

Pierce, R., & Stacey, K. (2004). Monitoring progress in algebra in a CAS active context: Symbol sense, algebraic insight and algebraic expectation. *The International Journal of Computer Algebra in Mathematics Education, 11*(1), 3–11.

Ruthven, K. (1996). Calculators in the mathematics curriculum: The scope of personal computational technology. In A. J. Bishop, et al. (Eds.), *International handbook of mathematics education* (pp. 435–468). The Netherlands: Kluwer Academic.

Soh, K. C., & Quek, K. S. (2001). Gender differences in TIMSS mathematics achievement of four Asian nations: A secondary analysis. *Studies in Educational Evaluation, 27,* 331–340.

Tan, H. (2005). *A comparative study of mathematics teachers' perceptions about computers and graphic calculators*. Unpublished master's thesis, Monash University, Melbourne, Victoria, Australia.

Tan, H., & Forgasz, H. J. (2006). Graphics calculators for mathematics learning in Singapore and Victoria (Australia): Teachers' views. In J. Novotná, H. Moraová, M. Krátká, & N. Stehliková (Eds.), *Mathematics at the centre* (Proceedings of the 30th Conference of the International Group for the Psychology of Mathematics Education, pp. 5-249-5-256). Prague, Czech Republic: Charles University.

Tan, H., & Forgasz, H. J. (2011). Students' ways of using handheld calculators in Singapore and Australia: Technology as master, servant, partner and extension of self. In J. Clark, B. Kissane, J. Mousley, T. Spencer, & S. Thornton (Eds.), *Mathematics: Traditions and [new] practices* (Proceedings of the 34th annual conference of the Mathematics Education Research Group of Australasia and the Australian Association of Mathematics Teachers). Adelaide: AAMT and MERGA.

Wong, K. Y., & Lee, N. H. (2009). Singapore education and mathematics curriculum. In K. Y. Wong, P. Y. Lee, B. Kaur, P. Y. Foong, & S. F. Ng (Eds.), *Mathematics education: The Singapore journey* (pp. 13–47). Singapore: World Scientific.

Chapter 6
Surveying the Public: Revisiting Mathematics and English Stereotypes

Helen Forgasz and Gilah Leder

Introduction

A fundamental goal of the mathematics curriculum is to educate students to be active, thinking citizens, interpreting the world mathematically, and using mathematics to make predictions and decisions about personal and financial priorities (ACARA 2009).

> The study of English… helps create confident communicators, imaginative thinkers and informed citizens. It is through the study of English that individuals learn to analyse, understand, communicate with and build relationships with others and with the world around them… It helps them become ethical, thoughtful, informed and active members of society. (ACARA n.d.)

From the quotations from the Australian Curriculum above, it is clear that mathematical and English language competencies are highly valued and integral to competent decision making and success in daily life. Mathematics continues to be an enabling discipline for Science, Technology, Engineering, and Mathematics (STEM) based university studies and related careers; English fluency is a necessary, and assumed, criterion in most employment contexts. It is therefore disturbing to find research indicating that males and females are still viewed differently:

> Negative stereotypes about girls' and women's abilities in mathematics and science persist despite girls' and women's considerable gains in participation and performance in these areas during the last few decades. Two stereotypes are prevalent: girls are not as good as boys in math, and scientific work is better suited to boys and men. (Hill et al. 2010, p. 38)

When it comes to English language studies, the prevalent gender stereotype is that females are better than males.

H. Forgasz (✉) · G. Leder
Faculty of Education, Monash University, Clayton, Australia
e-mail: helen.forgasz@monash.edu

G. Leder
e-mail: gilah.leder@monash.edu

A. Bishop et al. (eds.), *Diversity in Mathematics Education*,
Mathematics Education Library, DOI 10.1007/978-3-319-05978-5_6,

In searching for explanations for such typecasting it is instructive to look briefly at recent Victorian Certificate of Education (VCE) participation and performance statistics for grade 12, the final year of schooling in Victoria, Australia. This information is published by the Victorian Curriculum and Assessment Authority (VCAA). In 2011, 51,268 students were eligible to complete VCE. Of these, 49,835 completed VCE satisfactorily. The number of females exceeded the number of males both in enrolments (27,289 or 53 % and 23,979 or 47 %, respectively) and satisfactory completions (97.4 and 97.0 %, respectively). Data for two advanced and challenging VCE subjects, Specialist Mathematics and (English) Literature, are provocative. More boys than girls enrolled in the first of these two subjects (2671 and 1385, respectively), and more girls than boys in the second (4156 and 1697, respectively)—a startling sex[1] difference in the ratios of enrolments. In all VCE subjects, A+ and A are the highest grades awarded. In Specialist Mathematics and in Literature, a higher proportion of boys than girls obtained an A+ grade in all three assessment components (a school-assessed task and two external examinations) of these subjects.

With respect to achievement, it might appear that these data provide some evidence in support of the male stereotype in mathematics, but fail to support the female stereotype in English. Stereotypes, it would appear, are not always a reflection of reality, although beliefs about them are persistent. Stereotyping can have unwelcome spin-offs. Some people may assume that all members of the stereotyped group are similar, that is, they are all encumbered with the same negative, or positive, attribute. This theme is explored at some length in other chapters in this volume: interpretations associated with group rather than individual student achievement on large-scale tests are discussed in Chapter 2; issues raised by an assigned label such as "disengaged student" in Chapter 12. In this chapter, however, the focus is on the way members of the general public thought about boys and girls and mathematics and English. Were their views still gender stereotyped? And if so, does it matter—might prevalent societal views restrict the participation and achievement in mathematics, and English, of some groups?

Explanations for Gender Differences in Performance in Mathematics

In western countries, multiple explanations have been put forward for the persistent patterns of gender differences favouring males in mathematics achievement. After a detailed review of relevant literature, Halpern et al. (2007) concluded that the reasons for the overlap and differences in the performance of males and females

[1] In this chapter, we use "sex differences" when it is clear that categorisation is only based on biological factors. "Gender differences" are used when psycho-social or socio-cultural factors may contribute to any difference found.

could not be explained by a single factor, and that "[e]arly experience, biological constraints, educational policy, and cultural context" (p. 41) could all play a part. With reference to the USA, Geist and King (2008) similarly referred to pervasive societal beliefs about gender linked capabilities and their impact:

> Many assumptions are made about differing abilities of girls and boys when it comes to mathematics. While on the 2005 NAEP [National Assessment of Educational Progress] girls lag only about 3 points behind boys, this is only a recent phenomenon. In the 1970's, girls actually outperformed boys in all but the 12th grade test…. assumptions about differing levels of ability pervade not just the classroom, but home. (pp. 43–44)

In their detailed model of achievement motivation, and implicitly of academic success, Wigfield and Eccles (2000) highlighted the influence on students' learning and behaviours not only of learner-related variables, but also of the overall context in which learning occurs, that is the attitudes, actual and perceived, of critical "others" in the students' home and at school, and societal expectations more generally. That this is recognised by students themselves can be inferred from the comments made by one of the mathematically talented females interviewed by Leder (2011, p. 453): "An advantage of being a male would be to have been more encouraged to pursue a career in mathematics/engineering/technology. I would also have fitted in at high school better than I did". How often do students feel excluded because those around them expect conformity?

Through the media, gender stereotypes in mathematics can be reinforced (e.g., in print—see Forgasz et al. 2007) or challenged (e.g., on television—see Steinke 1998). Parents and teachers have been found to hold gender-stereotyped beliefs about and expectations of children's mathematical capabilities; these beliefs are often more strongly held by males (e.g., Tiedemann 2000). When it comes to perceptions of students' capabilities in English, parents are implicated. Bhanot and Jovanovic (2005) claimed that "when parents endorse the gender stereotype that English and social science are female domains, they tend to overestimate their daughters' abilities in these subjects and to underestimate their sons' abilities" (p. 597). In a now-dated study, Forgasz and Leder (1996) found that mathematics was perceived as a male domain and English as a female domain by both the high school boys and girls in their sample. Do these views persist to the present day? We turned to the general public to find out.

Additional Context for the Study

Mathematics and English Achievement

Annually since 2008, virtually all Australian students in years 3, 5, 7, and 9 complete National Assessment Program—Literacy and Numeracy (NAPLAN). A persistent pattern has emerged of boys outperforming girls in Numeracy but girls consistently outperforming boys in the reading, writing, spelling, and grammar and punctuation components of the test—see NAPLAN annual reports (downloadable from Australian Curriculum, Assessment and Reporting Authority [ACARA] website). Results from a large-scale international testing program, PISA 2009, revealed that in the reading literacy component, females outperformed males in all participating countries. In Australia the difference was 37 points, close to the Organisation for Economic Co-operation and Development (OECD) average (Thomson et al. 2011). Persistent gender differences in performance in aspects of English language performance have also been reported in the wider literature (Logan and Johnston 2009; Watson et al. 2010).

Generational Differences

The direction of generational differences reported in views on equity issues in the wider research literature are not necessarily as might be anticipated. It seems that younger adults are often more conservative than their older counterparts. In summarising results from several studies, Powlishta (2002, p. 169) claimed that "attitudes become more egalitarian with age", and that in their attributions of characteristics to males, females, or both/neither, "adults were less stereotyped in their attitudes than were children". In an examination of repeated cross-sectional surveys (1986–2005) of Australians' beliefs about family roles and men's and women's work, van Egmond et al. (2010) concluded that on most issues:

> Australian men and women have become increasingly more egalitarian in their views about gender arrangements. But the story is not so straightforward. The trends have taken a different direction since the mid-1990s.... Over the last 10 years attitudes to gender arrangements have shifted and the trend toward liberalization has slowed markedly and possibly stalled. (p. 162)

Evidence in support of older Australians being less gender-stereotyped in relation to mathematics learning was reported by Leder and Forgasz (2011) in an earlier study of the views of the general public and mathematics learning. Could a more negative outlook among the different generations of members of the public result in lower societal expectations for mathematics to be a field of study appropriate for all?

Societal Expectations: Public Views about Mathematics and English

Many explanatory models for gender differences in mathematics learning outcomes—achievement, participation, and attitudes—include the views of society at large (see Leder 1992). Interestingly, thoughts about the gender stereotyping of mathematics are less often gathered from the general public (Leder and Forgasz 2010) than from stakeholders such as students, parents, and teachers. It is rare in the research literature to find studies in which the general public's views about mathematics, or for that matter English, are reported.

More than two decades have passed since the *Maths Multiplies Your Choices* media campaign was run in Victoria, Australia. The focused aim of the campaign was to encourage parents to consider more carefully the impact that mathematics might have on their daughters' futures. It is intriguing that *Maths Multiplies Your Choices* was sponsored by the Victorian government department responsible for the labour force, and not the ministry overseeing education. A component of the campaign involved canvassing parents' attitudes towards their daughters' education and future careers (McAnalley 1991). When the campaign had run its course and was evaluated (Department of Labour and Mattingly Advertising 1989), it was deemed successful: awareness of the issue had been raised, parents were more positive about mathematics and science as careers for their daughters, and female enrolments in mathematics increased. Since that time, there have been no further surveys in Australia of societal views on issues associated with gender, mathematics, and careers. In the UK, Sam and Ernest (1998) noted that there were also few studies about the image of, and myths surrounding, mathematics. A similar lack of information was identified by Lucas and Fugitt (2007) in the USA. They used a ten-item survey to identify the views of residents in the mid-west of the country about mathematics, how it was taught in schools, and the effects of learning mathematics on young people's lives. We also explored these issues in the study reported in this chapter.

Canvassing Views from the General Public

Our study comprised three different, sequentially organised stages. In the first, we approached pedestrians in 12 different sites in Victoria—eight in the metropolitan area of Melbourne and four in regional/country Victoria—with a request to answer a short survey. Part of the explanatory statement prepared for the survey and made available to participants read:

> We have stopped you in the street to invite you to be a participant in our research study.… We are conducting this research … to determine the views of the general public about girls and boys and the learning of mathematics. We believe that it is as important to know the views of the public as well as knowing what government and educational authorities believe.

To broaden our sample beyond Victoria, we placed an advertisement with a link to an online version of the same survey on Facebook (see Forgasz et al. 2011 for details on how this was done). We targeted Australians aged over 18 (to meet research ethics demands) and thus gathered responses from people across Australia. Third, to provide a context for our findings about mathematics we placed another advertisement on Facebook linked to a comparable online survey about English. Overall, we received about 1000 responses to our surveys: over 650 from the pedestrians in the streets of Victoria, approximately 120 responded to the mathematics survey on Facebook, and over 160 to the equivalent English survey. In this chapter, we focus only on the data gathered via Facebook as we had not gathered data about English from pedestrians in the street.

Our Survey

To maximize cooperation and completion rates, the surveys about mathematics and about English were limited to 14 core items. The items focused on: the learning of mathematics/English at school, perceived changes in the delivery of school mathematics/English, beliefs about boys and girls and mathematics/English, their perceived facilities with calculators and computers, and their suitability for particular careers. Findings from 9 of the 14 items used on the survey are discussed in this chapter. By including two identical items on each survey (see 8 and 9 below), we had an inbuilt check whether the views of people who responded to one or other of the surveys were biased by their empathy for mathematics or English. The nine items are:

1. When you were at school, did you like Mathematics/English? [Yes/No]
2. Were you good at Mathematics/English? [Yes/No/Unsure]
3. Who are better at Mathematics/English, girls or boys? [Girls/Boys/Same/Unsure]
4. Should students study Mathematics when no longer compulsory/Should studying English be compulsory?
5. Who do parents think are better at Mathematics/English?
6. Who do teachers think are better at Mathematics/English?
7. Is it more important for girls or boys to study mathematics/to be good at English?
8. Who are better at using computers, girls or boys?
9. Who are more suited to working in the computer industry, girls or boys?

Some personal background information about respondents was also gathered: age (in terms of broad age-range), whether the respondents were male or female, and location (state) of residence in Australia. As well as readily codable responses such as *yes/no/do not know* or *boys/girls/the same/unsure*, respondents were encouraged to elaborate and explain the reason for their answers to the items. These elaborated responses were particularly informative.

Results

The Samples

The sample sizes, by gender and by age, of those who completed the two surveys—Mathematics and English—are shown in Table 6.1.

As can be seen in Table 6.1, younger respondents (that is those under 40 years of age[2]) dominated in both surveys. Facebook, it is known, has been embraced by

[2] Middle age is defined by both the Oxford and Collins dictionary as starting around the age of 40. We have therefore chosen 40, somewhat arbitrarily, as the age to discriminate between older and younger participants.

Table 6.1 The samples

Mathematics ($N=119$)			English ($N=161$)		
Male	Female	Unknown Gender	Male	Female	Unknown Gender
61 (51 %)	51 (43 %)	7 (6 %)	59 (37 %)	101 (63 %)	1 (1 %)
Under 40	Over 40		Under 40	Over 40	
98 (89 %)	14 (11 %)		123 (76 %)	38 (24 %)	

the younger generations as a popular means to socialise and share personal information. Interestingly, there were somewhat more male than female respondents to the survey about mathematics, and more female than male respondents to the survey about English. We wondered if this difference might be indicative of the (gender stereotyped) subject preferences of the general public, or at least of those who use Facebook.

For each item on the two surveys, we used chi-square tests to examine whether there were gender differences or differences by age group. Very few statistically significant differences were found by gender or age on either survey, no more than would be expected by chance. Although age was found to be a significant variable in our earlier study in which pedestrians answered the survey questions (Leder and Forgasz 2011), this was not the case with respondents on Facebook as the age profile was largely restricted to those under 40 (89 % for the Mathematics survey and 76 % for the English survey). In this chapter, we therefore report the results for the entire samples responding to each Facebook survey. Thus, as it turned out, the study reported here provided a sampling of the views of predominantly younger people and whether their views revealed strong stereotypes about mathematics and English.

In the next section, we report our findings using the relevant question(s) as the headings.

Questions 1 and 2

1. When you were at school, did you like Mathematics/English?
2. Were you good at Mathematics/English when you were at school?

Responses to questions 1 and 2 are shown in Fig. 6.1a and b.

From the graphs in Fig. 6.1a and b it can be seen that a majority of respondents indicated that they liked/had been good at mathematics and English when at school. The high proportion of those who had liked the subjects can be seen as a gratifying result for those of us involved in education. It is also worth mentioning that a higher percentage of respondents reported liking/being good at English than replied liking/being good at Mathematics. This, too, is worth noting.

Teachers and the scope of the curriculum featured strongly as contributors to both the like or dislike of each subject. Typical explanations included:

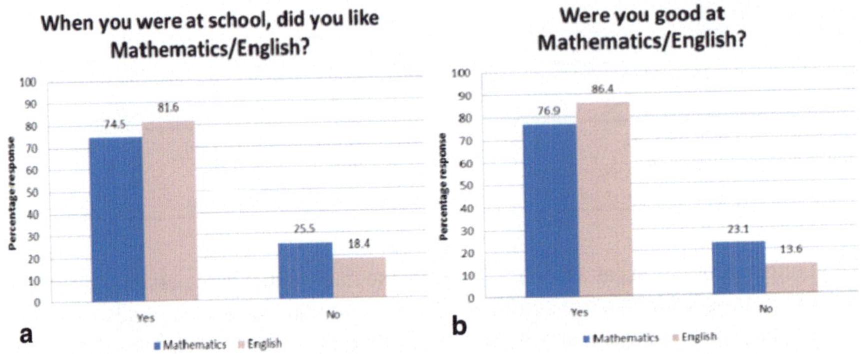

Fig. 6.1 **a** Response frequency to question 1 for mathematics. **b** Response frequency to question 2 for English surveys

> The teachers had wonderful methods of teaching us how to do the problems rather than just telling us to solve it. (liked mathematics)
>
> Found it very boring. Not a lot of time doing things in the real world. Teachers rarely inspiring. (disliked mathematics)
>
> In primary school and high school, I found mathematics challenging, but I did enjoy the challenge. I knew about the real world applications of the things I was learning, and the benefits of continuing to study maths, so continued with it. (liked mathematics)
>
> I excelled in English thanks to some fantastic teachers—all of whom created a stimulated learning environment. They taught me to appreciate some of the great literary works. (liked English)
>
> I greatly enjoy the complexities and intricacies of language. There is a pleasure in the manipulation and utilisation of language that can be derived from a particularly lovely turn of phrase, the use of idiom and double entendres and most importantly, the ability to convey emotion, colour and imagery. I also find grammar, syntax and the structure of language fascinating.... Poetry I find most attractive and learning about various techniques, structures and poets whilst I was at school was quite enjoyable for me. However learning how to write such literary works as expositions and biographies was also something that I enjoyed. (liked English)

Critical aspects of the aims of the Australian curriculum for mathematics and English learning, captured in the quotations at the top of the chapter, appear to coincide well with reasons participants gave for liking the subjects.

Question 3: Who Are Better at Mathematics/English, Girls or Boys?

Responses to question 3 are shown in Fig. 6.2.

As can be seen in Fig. 6.2, about half the respondents to both surveys indicated that they did not know or that they believed there were no differences between boys and girls in their achievement in mathematics or English. The responses of those who did consider that there are differences were startlingly gender stereotyped. For

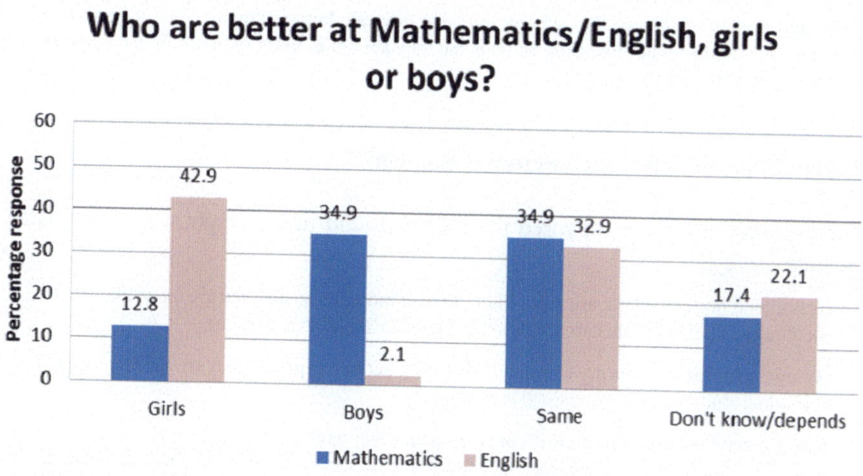

Fig. 6.2 Response frequencies to question 3 for mathematics and English surveys

mathematics, 35 % said boys and 13 % said girls; for English, 43 % said girls and 2 % said boys.

Open-ended responses are again shown below. Several themes dominated in support of identifying either girls or boys as better. These were: innate characteristics, social expectations, or personal experiences. These themes were also evident for many other questions on the surveys.

Explanations for "Who are Better at Mathematics"?

> Different part of brain used, I think! (boys—innate)

> Boys are given more confidence and it seems assumed that they'll understand more than girls. I would believe (there are) studies showing that male brains have a higher aptitude. (innate)

> Definitely more guys in the maths classes at university. (boys—experience)

> More girls than boys doing high level maths at my school. (girls—experience)

The last two of the above quotations reflect how different experiences can shape more general personal beliefs. The different answers—boys in one case, girls in the other—were both supported by personal histories.

The qualifications provided by some of those who answered "same" were also noteworthy. Socially appropriate expectations were often contrasted with personal experiences or beliefs.

> While I remember the boys in my maths class did better than the girls I still have a belief that marks are an individual consideration not one based on gender.

> Inherently, I don't think maths is a gendered skill. I believe a lot of girls are socialised to underachieve in maths.

Should be same but since the perception is that boys are better at maths, girls may choose not to take maths up because they believe that they are not as good so they self-select themselves out of the pool.

Explanations for "Who are Better at English"?

The explanations given mirrored the themes found in the responses to the mathematics survey.

Girls attend better in class, are more likely to read more for leisure and are more likely to be encouraged to excel in this area. (girls—experience/social expectation)

My own experience over my lifetime is that girls get better grades and are more interested in English studies. (girls—personal experience)

This is often the case because it is considered an acceptable past time for a girl to sit down and read a book (thereby gaining the experience necessary to excel at English) whereas boys are often looked down on for doing the same activity. (girls—social expectation)

Girls are socialised to express themselves more fully, therefore, they find it more natural to be creative in their writing at school. Also, boys are encouraged to engage in more active pursuits than reading. (girls—social expectations)

Girls are generally more academic and boys hands on, industrial. (girls—innate)

Girls are quite imaginative and analytically minded and tend to be able to think more laterally than their male counterparts. (girls—innate)

Question 4: Should Students Study Mathematics When No Longer Compulsory/Should Studying English be compulsory?

Responses to the slightly different versions of question 4 on the two surveys are shown in Fig. 6.3a and b.

As can be seen in Fig. 6.3a and b, a clear majority (71 %) of those who completed the mathematics survey believed that students should study mathematics when it was no longer compulsory, and almost all (92 %) responding to the English survey endorsed English being a compulsory subject. Reasons provided to support these views are shown below.

Explanations for Studying Mathematics When it is No Longer Compulsory

Young people must be able to work out figures for simple things like how much money they need to go out or pay rent even eat.

It's so vital for everyday life. I have a friend who was pretty good at maths in year 10, but dropped it in upper high school. She now can't even calculate change—she always thinks it's wrong when it's right, or she'll try to pay for something that costs $ 9 with $ 11, thinking she'll get a $ 5 note back.

I think it gives you logical thinking skills that can help you evaluate and solve problems in non-mathematical situations.

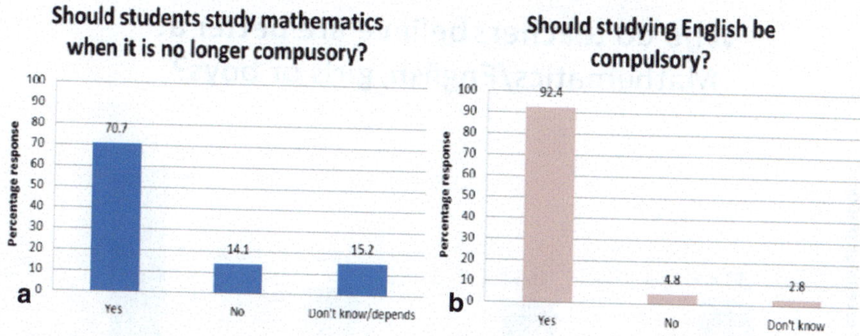

Fig. 6.3 **a** Response frequency to question 4 on the Mathematics survey. **b** Response frequency to question 4 on the English survey

Explanations for Studying English to Be Compulsory

> We live in an English speaking country and it is necessary to be able to communicate effectively.
>
> What is our world coming to if people cannot express themselves properly. OK, there is a vast range of what could be considered 'properly' but the English language has such a depth of words and meanings that the ordinary person simply doesn't use because they have never been exposed to it.
>
> It's our national language, we must learn to communicate in it and understand it to have a functioning society.
>
> Like learning English or history, it is vital for life as a skill.

Collectively, the reasons given for the need to study both mathematics and English are associated with the skills students need to be functional and contributing members of society. Implicitly, the comments reflect key elements of the aims of the Australian curriculum.

Questions 5 and 6

5. Who do parents believe are better at mathematics/English, girls or boys?
6. Who do teachers believe are better at mathematics/English, girls or boys?

The response patterns for questions 5 and 6 were virtually identical. The responses to the question about teachers (question 6) are shown in Fig. 6.4.

As can be seen in Fig. 6.4, the majority of respondents to both surveys did not know what teachers believed about this issue or believed that teachers would not differentiate between boys and girls. The responses of those who believed teachers would be biased (about one-third in both samples) are again strikingly different, but only for English. Over 30 % believed that teachers would think that girls were better than boys at English, compared with just 1 % who nominated boys. For mathematics no such difference was evident; 18 % of respondents thought teachers would say

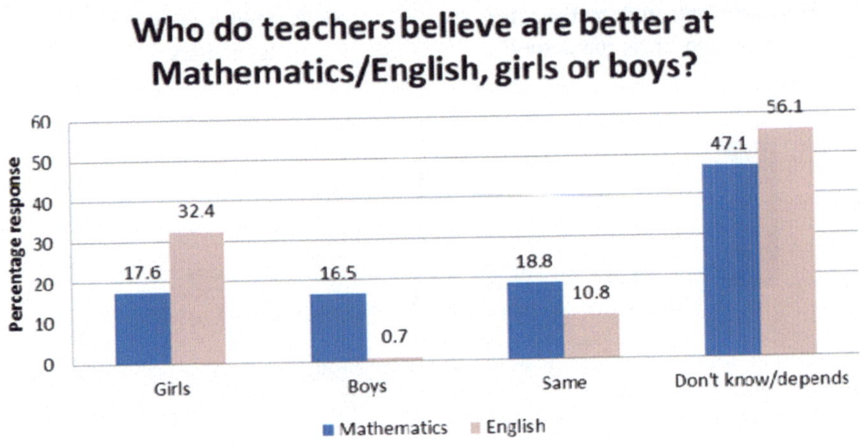

Fig. 6.4 Response frequencies to question 6 for mathematics and English surveys

that girls were better at mathematics compared to 17 % of respondents indicating that teachers would believe it was boys.

Explanations for Respondents Believing that Teachers Consider Boys to Be Better at Mathematics

> In my experience, teachers will be more inclined to aid mathematically gifted boys than girls of the same aptitude. (experience)
>
> I think many teachers may still feel that boys are naturally superior to girls in this subject, yet some may deliberately challenge this in their own practice and with the students directly or indirectly. (innate)

Explanations from Those Respondents Believing that Teachers Consider Girls to Be Better at Mathematics

> Possibly girls, but only due to the fact they are more mature and concentrate better in their teens. (innate)
>
> Going to a coed school I know my teachers knew that the girls in the maths classes were smarter—this is true though. They were able to concentrate and not get side tracked like us boys. (experience)

Explanations from Respondents Who Believed that Teachers Would Say that Girls are Better at English

> Because of the societal stereotypes surrounding English and the sexes. (social expectations)
>
> Because many (special) programs are initiated to help boys' learning. (experience)

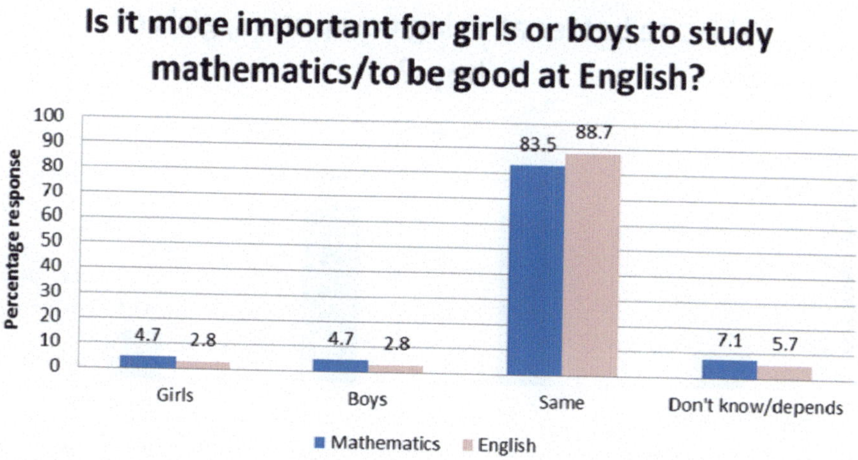

Fig. 6.5 Response frequencies to question 7 for mathematics and English surveys

> I've been a Maths teacher for 30 years. I have consistently seen more girls achieve higher marks in English than boys. (experience)

> Most likely because girls learn in different ways than boys and because of this are more attentive in class. (innate)

Similar explanations were given in answering the question about parents' beliefs about boys and girls and who would be considered better at mathematics and English.

Question 7: Is it More Important for Girls or Boys to Study Mathematics/to Be Good at English?

In the Australian context, mathematics is generally not compulsory beyond grade 10. However, historically the study of English has been required to grade 12 in some states/territories. Because of this, we used slightly different wording for this question on the two surveys. Responses to question 7 are shown in Fig. 6.5.

Figure 6.5 clearly reveals that the vast majority of respondents to both surveys said "same" (about mathematics: 84 %; about English: 89 %), that is, most respondents considered it equally important for both boys and girls to be proficient in both subjects. The more detailed, supportive, explanations provided invariably pointed to the importance of these subjects for both boys and girls—for careers and for everyday activities.

> The bias towards male superiority in mathematics must be vanquished. It is equally important for people of either gender to study mathematics, particularly if they have a strong interest in the field (whether it be pure, applied, stats etc). The generic skills developed by mathematics… are important to many aspects of life, for people of either gender. Employers value these skills very much, and so, the importance is equal. (mathematics)

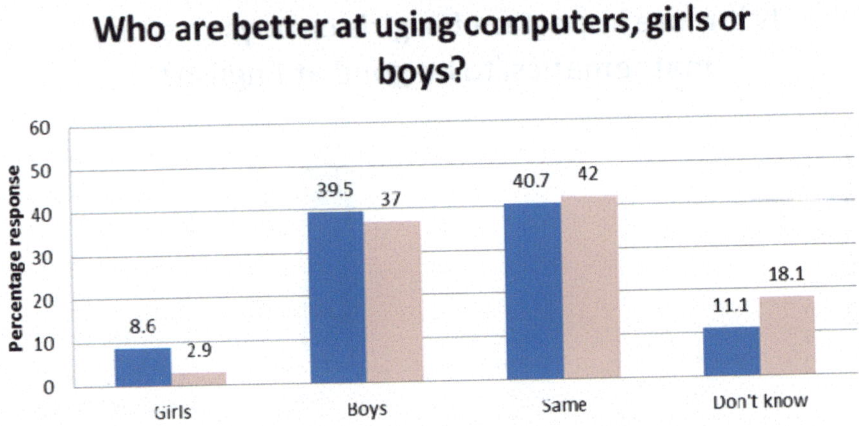

Fig. 6.6 Response frequencies to question 8 for mathematics and English surveys

> There is becoming less of a gap between "male" jobs and "female" jobs. There is now reason that it's more important for one gender than another. Should be same for anyone who wants a decent job. (mathematics)

> I can't imagine an argument that could favour one sex over another. (English)

> Everyone needs to be able to use English successfully. (English)

Questions 8 and 9: Questions Common to Both Surveys

8. Who are better at using computers, girls or boys?
9. Who are more suited to working in the computer industry, girls or boys?

The patterns of responses to these two questions on the Mathematics and the English surveys were almost identical. This gave us confidence that the surveys had been taken seriously and that the views expressed were common to different groups in the community.

The response patterns to question 8 are shown in Fig. 6.6. As can be seen in Fig. 6.6, a much higher proportion of respondents to both surveys said "boys" (mathematics survey: 40 %; English survey: 37 %) were better at using computers than said "girls" (9 and 3 %) were. It should be noted that a high proportion of respondents to both surveys said "same" (41 and 42 %), that is, that boys and girls are equally good at using computers.

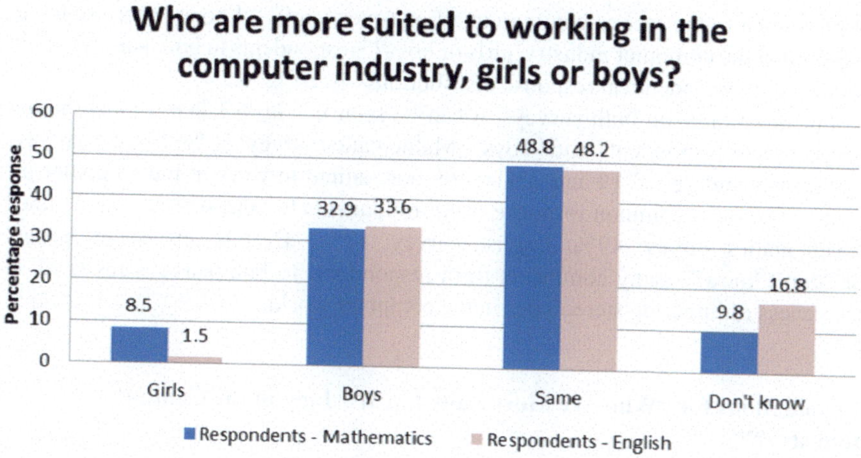

Fig. 6.7 Response frequencies to question 9 for mathematics and English surveys

Explanations for "Who Are Better at Computers, Boys or Girls?

BOYS: Maybe it is because they have more interest in computers. But this may have changed since I attended high school in the late 90's–early 2000's. (mathematics survey—experience/innate?)

BOYS: There is a perception (or maybe just me!) that, broadly speaking, boys are more interested in how things work and why they work the way that they do. (mathematics survey—innate)

BOYS: I've known more boys that can use computers well and they always seem to be able to solve all the problems (mathematics survey—experience)

SAME: There are various aspects of computing people may be accomplished in. For example, some girls may be highly talented in graphic art and design with Photoshop, which equates with being good at using a computer. Some boys may be highly talented in programming, again, demonstrating the capability to use a computer well. I don't believe that it makes sense to ask this question, due to the huge versatility of computers. Therefore, I've marked that they are equivalent. (mathematics survey)

BOYS: What I mean by yes, is that the median girl would be worse than the median boy. The best girls and boys with computers appear to be equal (I do IT at USYD and there are girls who are just as good as the boys). However there are many more boys who care enough about computers to become better. There is a lack of interest amongst girls and reliance on others. I think in an old fashioned sense, technology is the domain of the man of the house and girls have come to rely on their fathers and husbands. There is no reason why girls couldn't be as good, it's not like boys have some special talent. There are huge incentives to become good at using a computer if you were a girl (you should see the scholarships my uni offers to girls for IT). (English survey—social expectations)

With respect to the second common question (question 9) "Who are more suited to working in the computer industry, girls or boys?" respondents to both surveys again produced almost identical response distributions—see Fig. 6.7.

For question 9 on both surveys, it can be seen in Fig. 6.7 that a much higher proportion of respondents said "boys" (Mathematics survey: 33 %; English survey: 34 %) than said "girls" (9 and 2 %) were more suited to work in the computer industry. The most common response from respondents to both surveys was "same" (mathematics survey: 49 %; English survey: 48 %). Despite answering "same" or "don't know", many comments from respondents to both surveys revealed an awareness of lingering stereotypes in the computer world.

Explanations for "Who Are More Suited to Working in the Computer Industry?"

> SAME: Men are dominant in the computer industry, unfortunately. This has to be shifted back, while garnering more interest for women to pursue computer science/information technology. Either gender are (sic) equally suited to the field if they are talented enough and have a deep enough interest to pursue it. Gender inequality should not discourage, although it may be a choice factor in deciding not to work in the computer industry. (mathematics survey—social expectations)

> SAME: Immediately I thought boys, as this seems to be the current stereotype, but … more boys seem to be mathematically minded and therefore more boys undertake this kind of career, but the girls who are mathematically minded are just as suited to such an industry as the boys. (mathematics survey—innate, social expectations)

> SAME: I am in IT… there's no difference between genders… BUT the classes at college… out of a class of say 15, only 1 or 2 were women (English survey—experience)

> SAME: I believe that girls can do the same jobs as men however, the boys would be more socially accepted in the computer industry (English survey)

> SAME: I had years in the industry and it was WONDERFUL to get girls employed. In terms of stereotypes, they certainly seemed to have more understanding that they were there to service the needs of their clients in whatever the endeavour was, rather than to worship the great god of technology. But I'm not sure that it had anything to do with facility in English, just that their empathy seemed more directed towards people than machines. Competence both technological and personal was the ideal—and that can be combined in people of either gender. (English survey—experience)

> DON'T KNOW: Working in the computer industry may not just depend on particular skills but also on the culture of the profession/workplace. If you are entering a workplace with the demographics very different to you own, it can be difficult and you can be perceived as being unsuited for that type of work. For example, if the employees are almost all of the opposite gender or much younger or older age group. 'Suitability' is a culturally laden notion and very complex. (English survey—social expectations)

Concluding Comments

It is clear from our findings that most members of the general public in Australia who responded to our Facebook surveys value both mathematics and English studies. A large proportion is also unbiased in their views of the subjects as gendered domains. Regrettably, the views of those who do discriminate are consistent with the traditional stereotypes associated with mathematics and with English, that is, "mathematics is for boys, and English is for girls". Sadly, despite a paucity of biological evidence, innate characteristics of males and of females were identified by some as contributors to what becomes the obvious to these people. Personal experience also played a part for many participants. Observations of a particular context can, regrettably, lead to a propensity to generalise. The beliefs that evolve from personal experience become "the truth", and can be easily reinforced by media portrayals, for example. Consider the response of one female who disliked mathematics:

> [I] always achieved more in English than in mathematics. My older brother on the other hand was very good at mathematics at high school while his English was poor. We often helped each other to study for the subjects we did poorly in. (Female who disliked mathematics)

What conclusions about females and males was this woman likely to have made? It is tempting to infer that she will extrapolate from the particular to the general. We wondered how her parents may have responded to the situation, and what about her friends, and her teachers? On the survey she indicated that she did not know whether parents believed boys or girls were better at mathematics and gave no reason for this response. Teachers, she wrote would probably think girls were better, "although" she added "I remember the boys in my maths class did better than the girls…". To what extent do the views and opinions of others lead to exclusion from fields of study that might otherwise be of interest and challenge the intellect? Should one's biological sex be the constraint and determinant of a life direction? What does this say about a democratic society that cherishes independence, rewards achievement based on merit, claims to value mathematics highly, and claims to value diversity and inclusiveness? And, what is the effect on an individual, male or female, who challenges predominant stereotypes in society? Earlier research suggests that there is a price to pay if females deign to stray into the bastions of male dominance (e.g., engineering) such as (perceived) loss of femininity and eligibility for marriage and children; males, too, are often "labelled" if they enter traditionally female domains (e.g., nursing).

One of the more worrying features of the survey results was that the views expressed were those of younger (generally under 40 years of age) members of the general community—those who immerse themselves in the world of social media. As noted earlier, the younger members of our pedestrian sample who completed the mathematics survey (Leder and Forgasz 2011) were found to be more conservative in their gendered perceptions of mathematics than were their older counterparts, findings lending support to the generalised conclusions of van Egmond et al. (2010) about gender roles and stereotyping across Australia evidenced by changing trends since the 1990s. The Facebook samples were similarly less liberal in their views than we would have hoped. Has the realm of the social media and the immediacy and ease it presents to shape developing personalities and views had a role in this?

What are the implications of our findings with respect to mathematics curricula and the teaching of them? Over the years, there has been a plethora of intervention programs and recommendations to address gender bias aimed at promoting "gender inclusion" in mathematics and other STEM fields. Less attention appears to have been paid to the effects on boys of the gender stereotyping in the field of English. Teaching strategies have been proposed and modifications to curricular content and emphases put forward. Teachers and text book writers have been advised to take care that the selection of the contexts in which the mathematics to be learnt is exemplified are "gender inclusive" as well as revealing sensitivity to the diverse ethnic and cultural profiles of the nation. Since gender stereotyping of mathematics and English persists, questions remain about levels of awareness of these past issues, and their perceived relevance and importance in the challenging worlds of busy educationalists. A re-think is needed to avoid historical levels of gender stereotyping, and the consequences on a large talent pool, being repeated.

References

Australian Curriculum, Assessment and Reporting Authority (ACARA). (n.d.). Australian curriculum—English. http://www.australiancurriculum.edu.au/English/Rationale. Accessed 19 June 2014.

Australian Curriculum, Assessment and Reporting Authority (ACARA). (2009). Shape of the Australian curriculum: Mathematics. http://www.acara.edu.au/verve/_resources/Australian_Curriculum_-_Maths.pdf. Accessed 19 June 2014.

Bhanot, R., & Jovanovic, J. (2005). Do parents' academic gender stereotypes influence whether they intrude on their children's homework? *Sex Roles, 52*(9/10), 597–607.

Department of Labour and Mattingly Advertising. (1989). *Summary of two stage campaign evaluation study. Girls' career and subject choice*. Melbourne: Department of Labour and Mattingly Advertising.

Forgasz, H. J., & Leder, G. C. (1996). Mathematics and English: Stereotyped domains? *Focus on Learning Problems in Mathematics, 18,* 129–137.

Forgasz, H. J., & Leder, G. C. (2011). Equity and quality of mathematics education: Research and media portrayals. In B. Atweh, M. Graven, W. Secada, & P. Valero (Eds.), *Mapping equity and quality in mathematics education* (pp. 205–222). Dordrecht: Springer.

Forgasz, H. J., Leder, G. C., & Taylor, C. (2007). *Research versus the media: Mixed or single-gender settings?* Paper presented at the annual conference of the Australian Association for Research in Education. http://www.aare.edu.au/07pap/for07148.pdf.

Forgasz, H., Leder, G., & Tan, H. (2011). Facebook and gendered views of ICT. In S. Barton, J. Hedberg, & K. Suzuki (Eds.), *Proceedings of global learn Asia Pacific 2011* (pp. 1718–1727). Association for the Advancement of Computing in Education [AACE]. http://www.editlib.org/p/37393.

Geist, E. A., & King, M. (2008). Different, not better: Gender differences in mathematics learning and achievement. *Journal of Instructional Psychology, 35*(1), 43–52.

Halpern, D. F., Benbow, C. P., Geary, D. C., Gur, R. C., Hyde, S. H., & Gernsbacher, M. A. (2007). The science of sex differences in science and mathematics. *Psychological Science in the Public Interest, 8*(1), 1–51.

Hill, C., Corbett, C., & St. Rose, A. (2010). *Why so few? Women in science, technology, engineering, and mathematics*. Washington, DC: AAUW. http://www.aauw.org/learn/research/upload/whysofew.pdf.

Leder, G. C. (1992). Mathematics and gender: Changing perspectives. In D. A. Grouws (Ed.), *Handbook of research on mathematics teaching and learning* (pp. 597–622). New York: Macmillan.

Leder, G. C. (2011). Commentary 1 on *feminist pedagogy and mathematics*. In B. Sriraman & L. English (Eds.), *Theories of mathematics education: Seeking new frontiers* (pp. 447–454). Berlin: Springer.

Leder, G. C., & Forgasz, H. J. (2010). I liked it till Pythagoras: The public's views of mathematics. In L. Sparrow, B. Kissane, & C. Hurst (Eds.), *Shaping the future of mathematics education*. Proceedings of the 33rd annual conference of the Mathematics Education Research Group of Australasia (pp. 328–335). Fremantle: MERGA.

Leder, G. C., & Forgasz, H. J. (2011). The public's views on gender and the learning of mathematics: Does age matter? In J. Clark, B. Kissane, J. Mousley, T. Spencer, & S. Thornton (Eds.), *Mathematics: Traditions and [new] practices* (pp. 446–454). Adelaide: AAMT and MERGA.

Logan, S., & Johnston, R. (2009). Gender differences in reading ability and attitudes: Examining where these differences lie. *Journal of Research in Reading, 32*(2), 199–214.

Lucas, D. M., & Fugitt, J. (May 2007). *The perception of math and math education in the rural Mid West*. Appalachian collaborative center for learning, assessment, and instruction in Mathematics. Working Paper No. 37. http://www.eric.ed.gov/ERICDocs/data/ericdocs2sql/content_storage_01/0000019b/80/33/5b/c3.pdf.

McAnalley, K. (1991). Encouraging parents to stop pigeon-holing their daughters: The "Maths multiplies your choices" campaign. *Victorian Institute of Educational Research Bulletin, 66*, 29–38.

Powlishta, K. K. (2002). Measures and models of gender differentiation. In L. S. Liben & R. Bigler (Eds.), The developmental course of gender differentiation: Conceptuality, measuring and evaluating constructs and pathways. *Monographs of the Society for Research in Child Development, 67*(2), 167–178.

Sam, U. C., & Ernest, P. (28 Feb 1998) *A survey of public images of mathematics*. Paper presented at British society for research into learning Mathematics. http://www.bsrlm.org.uk/IPs/ip18-12/index.html.

Steinke, J. (1998). Theory into practice: Connecting theory and practice: Women scientist role models in television programming. *Journal of Broadcasting & Electronic Media, 42*(1), 142–151.

Tiedemann, J. (2000). Parents' gender stereotypes and teachers' beliefs as predictors of children's concept of their mathematical ability in elementary school. *Journal of Educational Psychology, 92*(1), 144–151.

Thomson, S., De Bortoli, L., Nicholas, M., Hillman, K., & Buckley, S. (2011). *Challenges for Australian education: Results from PISA 2009*. Melbourne: ACER. http://www.acer.edu.au/documents/PISA-2009-Report.pdf.

van Egmond, M., Baxter, J., Buchler, S., & Western, M. (2010). A stalled revolution. Gender role attitudes in Australia, 1986–2005. *Journal of Population Research*. http://www.springerlink.com/ content/83t4k56g58866h43/fulltext.pdf.

Watson, A., Kehler, M., & Martino, W. (2010). The problem of boys' literacy underachievement: Raising some questions. *Journal of Adolescent and Adult Literacy, 53*(5), 356–361.

Wigfield, A., & Eccles, J. S. (2000). Expectancy-value theory of achievement motivation. *Contemporary Educational Psychology, 25*, 68–81.

Chapter 7
Commentary for Section 1: Linking Research and Practice in School Mathematics

M. A. (Ken) Clements

What, then, do the chapters in this volume add to the extensive literature that already exists on the subject? The answer is, simply, "an enormous amount," because of the authors' experience as teachers, internationally recognized skills and experience as collaborative researchers, and much individual and collective wisdom. There is a wholesome unity about the section—the chapters are all concerned with describing conditions as they exist and then bringing pressure to bear on those who might be able to do something about making school mathematics more inclusive. The research summarized in this section was carried out in all states of Australia, in China, in Singapore, and in the USA.

Chapter 1 is written by Peter Sullivan, who has had large experience teaching and researching in schools and universities in Australia and Papua New Guinea, and most recently has been advising the Minister of Education in the Northern Territory of Australia on how best to support the teaching of literacy and numeracy in that State—and in particular on "the challenges of converting research findings to useful and practical advice." Sullivan discusses, with typical candor, the need to minimize factors that are likely to inhibit the learning of particular groups. The chapter is especially important for readers planning to teach or to conduct research associated with Aboriginal students. Sullivan recommends the inclusion of practitioners in research teams, and emphasizes the need for researchers to articulate implications for practice.

Gilah Leder (Monash University) and Sarah Lubienski (the University of Illinois at Urbana-Champaign) combined for Chapter 2, which is mainly concerned with large-scale test data deriving mainly from the National Assessment Program—Literacy and Numeracy (NAPLAN) in Australia and the National Assessment of Educational Progress (NAEP) in the USA. The chapter also reports and interprets data generated by other assessment programs such as Trends in Mathematics and Science Study (TIMSS), the Programme for International Student Assessment (PISA), and the (American) Early Childhood Longitudinal Study (ECLS). Leder

M. A. (Ken) Clements (✉)
Illinois State University, Normal, USA
e-mail: macleme@ilstu.edu

A. Bishop et al. (eds.), *Diversity in Mathematics Education*,
Mathematics Education Library, DOI 10.1007/978-3-319-05978-5_7,

and Lubienski's analysis reveals that, although since 2008 NAPLAN numeracy scores have remained steady, national reading spelling and grammar scores have been improving. Furthermore, the performance of Indigenous students, as a group, on the numeracy test remains well below the scores of other groups of students.

Leder and Lubienski make some clear recommendations for achieving improvement:

1. Practitioners, curriculum developers, and education systems managers should set clear instructional targets for curriculum areas in need of special attention.
2. Advantage should be taken of the "teaching-to-the-test" tendency by testing non-routine problem solving, reasoning, and writing solutions.
3. Assessment instruments should be broad enough to deter those who seek to narrow the curriculum so that "students will do well on the test".
4. Mathematics education researchers should make sure that on matters affecting school mathematics they—and not researchers outside the field of mathematics education—control research designs and instrument development.
5. Researchers need to pay much more attention to when gaps based on race, socio-economic status (SES), and gender begin, and to the circumstances under which they widen or narrow. Furthermore, any suggestion that such gaps are inevitable and fixed needs to be countered.

It is interesting that, whereas the large majority of Australian students in Years 3, 5, 7, and 9 participate in NAPLAN testing, in the USA, relatively small representative samples (less than 5 % of all students at the appropriate age) participate in NAEP tests. Given that statewide testing and/or national testing in Australia has been occurring for many years, now, and given that since the mid-1990s there have been state or national mathematics curricula in place, with associated statewide or national testing, one can question whether the NAPLAN scheme of testing most students is based on sound principles. Is it pure coincidence that performance of Australian students on TIMSS, relative to the USA, and other Western nations, has declined during the time that statewide or national testing has been more or less enforced?

The relatively strong performance of Australian students on PISA is also a matter of interest. Australia has had, for several decades now, an impressive group of researchers with strong teaching experience and close links to schools. There has been an abundance of rich tasks developed, and major state-supported programs have not only engaged teachers actively in the development of such tasks but also in the professional development of teachers to take full educational advantage of the tasks (Clements 2008). I believe that has had, and continues to have, a strong positive effect on the problem-solving performances of Australian students.

The third chapter, by Debra Panizzon, focuses on the impact of geographical location on student achievement. Panizzon reported PISA and other data showing that in Australia metropolitan students significantly outperform provincial students who significantly outperform students living in remote parts of Australia. Part of this is explained by the fact that non-Indigenous students significantly outperform Indigenous students. Panizzon worked with the National Centre for Science, Information and Communication Technologies and Mathematics Education for Rural and Regional Australia when that Centre collected data on, among other things, teacher attraction and retention. It was found that metropolitan zone schools had, relatively speaking, much less difficulty than remote zones in attracting and retaining qualified mathematics teachers, and that many secondary mathematics teachers in remote schools were not only unqualified to teach mathematics but were forced to teach composite classes in which students at different grade levels were placed. Panizzon stressed the need to counteract such effects on low SES and Indigenous students, and her recommendations should challenge all systems of education within Australia.

Tasos Barkatsas and Wee Tong Seah's chapter on learners' preferred mathematical task types, and the values to be associated with those preferences, summarizes intriguing research comparing preferences and values, with respect to mathematical tasks, of fifth- through eighth-grade students in Victoria and fifth and sixth graders in two very large cities (Chongqing and Chengdu) in China. It was found that, although student preference for specific types of mathematics tasks varied across geographical locations, the students in each of the three research settings tended to value, highly, real-life scenarios, questions that were challenging but within-reach, and easiness. That finding was true for questions for both number and geometry topics. Barkatsas and Seah claimed that this finding represents "a strong message for efforts that facilitate inclusiveness for all students." It was especially interesting that students in Chongqing and Chengdu, which are located only 300 km from each other, did not have the same preferences.

Hazel Tan's chapter on the preferences of 964 senior secondary students in Singapore who were learning to use graphics calculators and calculators with computer algebra systems provides an important reminder to all concerned with mathematics education that the world of mathematics education is changing very rapidly, and technology is a major factor influencing the directions of change. One might have thought that Singaporean students' well-known high performances on TIMSS and PISA would have made those directing school mathematics education policies in Singapore very reluctant to embrace graphics calculators and computer algebra systems, but Tan, who has had much teaching experience in the Singapore system, makes clear that the reverse has been the case. Nevertheless, Tan's analysis indicated that there were gender-related differences in the ways Singaporean students preferred to use graphics calculators, and that these differences were not always recognized or taken account of by teachers of mathematics.

Singapore education authorities have not only recognized that technological developments are challenging traditional views on curriculum, teaching, learning, and assessment, but they are also making policy decisions aimed at achieving a greater degree of inclusiveness. How can busy teachers keep up with developments and help increasing numbers of students to understand mathematics well? Singapore has faced such questions squarely and, like the state of Victoria, has bravely determined to allow students to use graphing calculators in high-stakes university-entrance examinations. After drawing attention to gender-related differences, Tan concluded that "teachers planning gender inclusive practices need to take into account a host of factors including students' confidence and competence with the calculators, attitude and competency in mathematics, and learning preferences." What is needed, she emphasizes, is a "student-centred pedagogy that promotes quality learning of every individual, rather than teaching to different 'labels' such as gender."

In the sixth and final chapter of this section, Helen Forgasz and Gilah Leder, both internationally recognized researchers on gender differences in mathematics education, took advantage of opportunities presented by Facebook to investigate the attitudes of 120 adults who responded, online, to an "attitudes to mathematics" survey and 160 who responded, also online, to a similar "attitudes to English" survey. Forgasz and Leder concluded that, although many did not gender stereotype mathematics or English, among those who did, "the traditional stereotypes exist." My own interpretation of the data, as it was reported, was that it was encouraging to find that well over 80 % of adults, in each group, said they thought that boys and girls were equally good at mathematics, and the same was true for English. Almost half the adults in each group reported that girls and boys were equally suited to working in the computer industry, and the number who thought girls were more suited was about the same as the number who thought that boys were most suited. These and other data reported by Forgasz and Leder suggested to me that gender stereotyping with respect to mathematics, English, and computer education has decreased sharply over the past two decades.

References

Clements, M. A. (2008). Australasian mathematics education research 2004–2007: An overview. In H. Forgasz, A. Barkatsas, A. Bishop, B., Clarke, S., Keast, W. T., Seah, & P. Sullivan (Eds.), *Research in mathematics education in Australasia 2004–2007* (pp. 337–356). Rotterdam: Sense Publications.

Forgasz, H., & Rivera, F. (Eds.). (2012). *Towards equity in mathematics education: Gender, culture, and diversity*. Berlin: Springer.

Jablonka, E., Wagner, D., & Walshaw, M. (2013). Theories for studying social, political and cultural dimensions of mathematics education. In M. A. Clements, A. J. Bishop, C. Keitel, J. Kilpatrick, & & F. Leung (Eds.), *Third international handbook of mathematics education* (pp. 41–68). New York: Springer.

Part II
Interrogating the Boundaries

Chapter 8
Diversity, Inclusion and Equity in Mathematics Classrooms: From Individual Problems to Collective Possibility

Mike Askew

Introduction

'We think, hear and speak reality into action'. (Carrington and MacArthur 2012)

Diversity, differentiation, difficulties: the literature on mathematics education is replete with 'd-words'. How we think and talk about diversity impacts not only in terms of how we react to the differences that learners actually bring to the mathematics classroom but also upon what differences are brought into being through talking and consequently acting. In other words, while there are, and always will be, differences between learners, a key question we have to ask is to what extent are learner differences in mathematics education a result of such natural variation or to what extent they are 'spoken into reality', are a social creation?

In this chapter, I argue that the discourse in some schools, classrooms and policy circles frames diversity as a barrier to effective teaching or learning—diversity must be either reduced, through practices like setting and streaming, or it has to be 'managed' to reduce its impact, through practices like individualised learning experiences. These discourses of reducing or managing diversity arise, in part, through talk about diversity being framed around the causes of the 'problem' of diversity being 'located' primarily within the individual learner. Such a position on diversity rarely, however, questions assumptions about the mathematics curriculum. The curriculum is taken as a given, it is the learners or the teachers who have to find ways to accommodate to that given. Individual learners have to be helped to engage more effectively with a predetermined curriculum or pedagogies developed to include more individuals become the focus of inquiry. In a climate of increased specification of curricula intentions, THAT learners should be expected to engage with THAT curriculum become unquestioned givens. Thus, the traditional, Western-type approach to education and current schooling and classroom structures where one teacher is expected to bring groups of students to one-size-fits-all outcomes turns diversity into a problem. Given that such systemic constraints are unlikely to change in the

M. Askew (✉)
Faculty of Education, Monash University, Clayton, Australia
e-mail: mike.askew@monash.edu

A. Bishop et al. (eds.), *Diversity in Mathematics Education*,
Mathematics Education Library, DOI 10.1007/978-3-319-05978-5_8,

near future, in this chapter, I suggest that rather than trying to re-frame diversity directly, a reconsideration of how we think about curriculum could have a positive impact on student inclusion.

Thus, rather than taking the curriculum as a given, so that ways have to be found to help teachers enact and learners attain that curriculum, I turn this around by looking at assumptions about the curriculum and how these may be 'speaking' diversity into being a problem. I then consider different ways of thinking about the curriculum and how these different ways may have practical implications for more inclusive and equitable mathematics education.

In particular, I argue that a more productive stance towards curriculum focuses on the classroom collective and working with the collective construction of mathematical knowledge, in contrast to the commonly held position of trying to meet the needs of each individual learner and practices that flow from that position. Rather than taking the individual as the starting point for planning learning experiences I argue that practices starting from the position of building learning communities are more inclusive while still ultimately addressing the needs of the individuals within that community. Achieving the equitable goal of removing 'barriers to learning whilst challenging each student take risks and responsibility for learning' (Small 2012) is thus achieved through creating collective classroom cultures that support the individual learner.

It is not, however, my intention to replace one set of dogma ('learner-centred teaching') with another ('teach to the collective'). A focus on the collective should not, cannot, neglect the needs of the individual. But taking a stance towards curriculum that has at its centre the building of learning communities through dialogue can, I argue, be a disciplined approach to the issues of equity and inclusion.

Diversity and Equity

Ladson-Billings (1997) reminds us that any work on equity in mathematics education should not seek solutions to the issues of differential attainment without first developing a clearly articulated theoretical position on what actually are the problems and issues underlying equity and diversity. My starting point, therefore, is to briefly set out my position on issues of diversity and equity.

Education in a representative democracy is an inherently and inevitably political enterprise: teaching cannot avoid enacting particular visions of desirable learning outcomes (Reid and Valle 2004). Any education system is selective in what it values and no system can avoid excluding some students. But what is unacceptable is to consistently and continually exclude certain groups of students. In Australia, for example, Indigenous, rural and low socioeconomic status (SES) groups consistently show lower success rates on outcome measures valued by the system—national test results and rates of continuing in education (see, for example Panizzon Chapter 3).

The achievement of equity in mathematics education would mean that it became impossible to predict a student's achievements on the basis of characteristics such as which cultural groups they belong to (Gutiérrez 2007). Mathematical achievement should be no more predictable from the basis of knowing whether a student is male or female, Indigenous or Asian, hearing impaired or not, than on knowing whether they prefer coke or Pepsi, wear sneakers or boots, support Newcastle or Manchester. In an equitable system, the natural variation between people would be reflected in a similar range of variation existing within (cultural) groups, rather than, as is currently the case, between these groups. The challenge for equitable mathematics education is for no particular group of students to be privileged.

> I would expect teachers over the course of their careers to find ways to counter the ability of any single group of students to command power in the classroom. Teachers should be able to look at their classes and not be able to see predictable patterns of achievement (e.g., on standardised tests, weekly exams, mastery of mathematical discourse) throughout a given year or across years. (Gutiérrez 2007, p. 43)

We should thus not confuse equity in terms of justice or fairness with the notion of equality in terms of sameness. Diversity does exist, will continue to exist in mathematics classrooms, and it is naïve to think that diversity can ever be completely eliminated (let alone whether this would actually be a desirable outcome). To accept that diversity will always be present is to acknowledge that people vary in their goals, interests and strengths and that any attempt to reach equality of outcomes, aside from being impossible, 'does not represent "justice" for students' own desires or identities' (Gutiérrez 2007, p. 41).

But accepting the natural diversity that comes about through individuals' different strengths and interests must not stop us questioning and challenging the 'normed' diversity that arises through practices such as labelling students or groups of students—practices that may actually create the very differences that such practices seek to reduce. Before turning to argue how rethinking the curriculum may be a positive move in the direction of equity, I briefly consider the implications of the emphasis of locating the 'problem' of diversity within the individual learner.

The Focus on the Individual Learner

The disengaged learner, the special needs learner and the gifted learner—how do such labels construct the learner and what is their impact? Difference may not be socially constructed but labels are socially and culturally constructed and although the labels applied to learners may change over time, issues of exclusion and lack of equity may be perpetuated through labelling practices. In California, for example, learning difficulties used to be defined as falling into three, hard to define, categories of special needs: mental retardation, learning disabilities and emotional disturbance. Eliminating the category of mental retardation did not, however, have any impact on the proportion of learners subsequently classified as having special needs: as numbers in the 'mental retardation' group went down, so numbers in latter two went up (Kavale and Forness 1998). Did these labels really identify needs or did they simply provide a way to bracket off learners who did not fit with expected norms?

Schooling 'is filled with instructions for coordinating the mutual construction of success and failure. … Categories such as "low achiever" or "learning disabled" are positions in education that get filled by children' (Varenne and McDermott 1999, p. 152). Varenne and McDermott highlight the importance of rather than assuming people fall into natural groups and consequently have particular, fixed, attributes, we need to see schooling as a system that dynamically produces these groups. Describing someone as being, say, a low achiever is not the same as saying they have green eyes—there is no inherent attribute underpinning being a low achiever—or a high achiever for that matter—both are a consequence of the dynamics of the system. Much of our labelling of learners and their attributes is, essentially, arbitrary.

Labelling, however, reinforces our everyday sense that categories of learners exist in a reality that lies beyond labelling practices. To reiterate: this is not a denial that diversity exists—some learners do have difficulty with mathematics and struggle in mathematics lessons. But all learners are unique and labelling a student as, say, a low attainer in mathematics does not mean that they are part of a homogeneous group—many different behaviours or circumstances can lead to low attainment. Any learner may struggle at some point in time but that should not, along with other categorisations, bring into being the assumption that they will always struggle (indeed without struggle we might argue that learning cannot occur).

Labelling itself is not the problem, but the subsequent unequal assigning of value is problematic. Take, for example, the values attached to labelling some learners as engaged or disengaged. To take an instance outside schooling, I am quite disengaged with cricket but it is only in some situations (mainly in a bar!) that this is a problem and even then I can often side step getting into conversation about it. Similarly, if mathematics were not a compulsory school subject, then disengagement would not be a problem. It is not that learners are predisposed to be engaged or not with mathematics, it is that they are forced into a situation where on structural (compulsory schooling) grounds being engaged is taken as the norm and disengagement the problem—and one that most often lies in the individual. Unmotivated, lack of motivation—the language of disengagement is one of something missing, or of an active uncoupling, further suggesting that the fault lies in the individual learner. Mathematics is unlikely to become an optional subject, but recognising that learner disengagement is, to a large extent, a consequence of the political decision to make mathematics compulsory frames the situation differently.

One counter argument here is that results of international rankings such as TIMSS and PISA have shown that entire nations can do well in mathematics (or at least do well on tests) and, assuming that the natural variation within these nations must be no different from anywhere else, then surely, the argument goes, anyone must be able to learn mathematics. If Finland/Singapore/South Korea/(insert your favourite 'high performing nation') can 'do well', then for any nation not ranking as highly the difficultly must lie either in the individual learners' attitudes rather than abilities (setting aside the claims that teaching in high-performing countries must somehow be more effective).

Such a shift of emphasis away from ability to attitude is mirrored in a rise in popularity of psychological accounts suggesting that succeeding or not at mathematics is down to a particular mindset and the importance of individual effort (Dweck 2000). The downside to narratives of mindset or effort is that while they shift the cause—from innate fixed ability accounts to malleable beliefs or practices—they do not shift the location of the difficulty away from the individual student. (To be fair, researchers in these areas do not attribute 'blame' in this way, but accounts drawing on the research often have a tacit sense of this.) Students do not succeed because they lack the will power or resilience to put the effort into mastering the mathematics. This is not to suggest that research into mindset or effort is not important and it does help account for some different learning outcomes, but it can perpetuate the view that what needs 'fixing' is the individual.

Further to this, the research into Pacific Rim nations that do well in international tests suggests, however, that many other features may be at play in these nations' successes: the extensive after-school cramming for students; the peer pressure to keep up, the 'honour' to one's family in succeeding and the highly competitive entry to university (Askew et al. 2010). Simply exhorting students outside that cultural milieu to try harder is not only unlikely to raise standards but also to reinforce individuals' views of themselves as failing. And linked to this are current shifts in the way curriculum is constructed by policy makers.

Policy and Curriculum

I start my journey through models of curriculum with the now classic formulation of the curriculum as intended, implemented and attained (Robitaille and Dirks 1982). While this model may now be so commonplace as to be noncontentious I argue that, largely as a result of wide-spread national and international assessment practices, these three dimensions of curriculum are becoming conflated: policy drivers aimed at reducing the spread of learner outcomes has led to the attained curriculum becoming perceived as being matched to, and enshrined in, the intended curriculum.

Prior to Robitaille and Dirk's tri-partite definition, curriculum was typically discussed in terms of the intended, with this sometimes reduced to a syllabus. For example, in the 1990s, in England and Australia, curricula were determined either through local authority or state guidelines or established post facto through teachers following the content of textbooks or the syllabi produced by exam boards.

Within such policy contexts, Robitaille and Dirk's model provided a helpful way of theorising and examining how the curriculum not only had an intended (even if not explicitly articulated) dimension, but also how this would necessarily be different from what students learned—the attained curriculum. Between the intended and attained the importance of mediating effects of teaching—the implemented curriculum—were thus foregrounded.

More recently, studies and trends have contributed to shifts in the curriculum landscape. The availability of copious local, national and international data on the attained curriculum (as measured by test scores) and the national anxiety (according to the media at least) about a country's ranking in international league tables means that the discourse has shifted away from accepting that there will be variation in the attained curriculum to the expectation that all learners (or at least the vast majority) can, and will, attain certain expected learning outcomes. The intended and the attained curriculum are increasingly conflated.

This shift is not without benefits. It challenges deficit models of populations: students from socio-economic backgrounds or cultural groups that historically have shown patterns of low attainment have been demonstrated, in some schools at least, to have standards of attainment that match or even exceed those of schools with more 'favourable' intake (as in coming from backgrounds that predispose students' to the schooling's model of teaching and learning). Deficit views of learners have been exposed: low attainment can no longer be excused on the grounds of 'these children'.

But there is also a downside to the conflation of the intended and attained curricula. What is to be learned is now perceived, in policy circles at least, as able to be clearly prespecified. In England, for example, the first national curriculum in the early 1990s set out 'programmes of study' that outlined what students were entitled to be taught. Learning outcomes were set out separately in 'levels of attainment'. This distinction between input (intended) and output (attained) was a result of review of research findings (Task Group on Assessment and Testing (TGAT) 1988) that not all learners might reach the same levels of attainment, with there being up to a 7-year 'gap' in attainment in mathematics at age 11 (Cockcroft 1982). Levels of attainment would be used to monitor individual learner's growth in attainment rather than set benchmark levels for particular ages. At the time of writing this chapter, over two decades later, the most recent iteration of England's national curriculum for mathematics no longer has any separation of the intended from the attained: all curriculum intentions are framed as what children will be taught to do, rather than what they are entitled to meet.

Similarly, the Australian National Curriculum provides a de facto intended curriculum through the adumbration of learning outcomes (attained). In both nations, the curriculum is now set out in terms of annual expectations of learning outcomes, thus creating norms around rates of learning. Deviation from these norms then has to be accounted for, with learners whose progress falls outside these norms, especially those making progress 'more slowly' than expected, seen as presenting challenges.

The conflation of the intended and the attained raises questions about the nature and position of teaching—the implemented curriculum. Pedagogy is effectively reduced to a conduit between the intended and attained, fostering a technical-rationalist approach to teaching and learning: select a predetermined, atomised, learning outcome, teach to that outcome and test for 'mastery'. Mathematics is reduced to skills-based behavioural objects and the irony is that teachers found to be 'teaching to the test' are criticised for this practice.

On the other hand, teachers are still expected to attend to the needs of individual learners, and since not all learners will be able to engage equally with prespecified learning outcomes, the practices of grouping learners between or within classes is perpetuated, as different learning objectives can then be set for different groups' 'needs'. As the research clearly demonstrates, such practices over time actually increase the gap in attainment, and the cycle of inequity is perpetuated (see for example, Wiliam and Bartholomew 2004; Oakes 2005).

Are there other, more constructive, ways of thinking about curriculum that might challenge current practices and encourage pedagogies that may be more equitable? I now examine this question by examining other models of curriculum.

Curriculum: Fact, Activity or Inquiry?

An alternative model for thinking about curriculum that is productive in terms of diversity and inclusion comes from Smith and Barr (2008) who suggest thinking about:

- Curriculum as fact;
- Curriculum as activity, and,
- Curriculum as inquiry.

Curriculum as fact, they argue, views the link between teaching and learning as non-problematic: learning is simply a result of what is taught. In the classroom the teacher is the expert and learners consequently cast into relatively passive roles.

The idea of curriculum as fact is widely criticised in the mathematics education literature, and it would be unfair to ascribe such as view to many teachers. Policy directions, however, as argued above, can be read as treating the curriculum as fact and thus carrying implicit messages about the relationship between teaching and learning as a straightforward one of transmission. The discourse of, for example, expected annual learning outcomes effectively talks into being a set of 'facts' (or skills or procedures) that, with appropriate teaching, should present few difficulties for the majority of learners the majority of the time. Curriculum as fact fits with views of the intended and attained curriculum as coterminous and the positioning of the individual or the teaching at the root of the 'problem' of differential outcomes.

An improvement on curriculum as fact is to think of curriculum as activity, broadly based on constructivist principles (Von Glasersfeld 1990) of learning and positioning learners as active sense-makers engaged in constructing meaning and connections. The teacher's role now becomes one of facilitator, structuring lessons so that individuals maximise the potential of making sense of their classroom experiences in the (mathematical) ways intended. While this moves in the direction of equity, the emphasis, in Smith and Barr's opinion, is still on the individual learner as the primary unit of learning.

Curriculum as inquiry goes one step further in taking a stance towards knowledge as co-constructed and emerging through interaction with others. Importantly, curriculum as inquiry shifts the focus away from the individual learner and onto the activity of the collective. This shift means that while there are still learning gains for the individual these go beyond acquiring knowledge: through becoming part of a community (knowing), learners also learn to learn (Smith and Barr, op cit).

A view of curriculum as inquiry still has a place for an intended curriculum, but the coupling between the intended and attained curriculum becomes looser. The intended curriculum, for example, is often spoken of, metaphorically, as a map that can be used to plot out and plan a prespecified journey on which to take learners. Curriculum as inquiry views the map more like a sea chart in sailing; a destination is chosen but the journey is subject to unpredictable and uncontrollable contingencies. The sea chart is used to monitor progress—it cannot be used to completely predetermine the route to be taken. Curriculum as inquiry has ends in mind but recognises that the route to these is not highly predictable, and that sometimes, a different destination may be preferable.

Curriculum as inquiry still has to be prepared for, even if lessons cannot be planned in the finest detail. This distinction between planning and preparing is not simply one of semantics. Being prepared for what may emerge in a lesson is not the same as planning for what will be taught. Saying that lessons should not have predetermined goals is not the same as saying lessons should be directionless. Preparation means teachers can build on what learners bring to the lesson while still keeping an eye on the mathematical 'horizon' (Ball et al. 2005). As the skilled sailor responds to conditions, so the skilled teacher has to be responsive to what happens in the lesson and both need to adjust and act to keep moving towards their destinations.

A difficulty with over-planning activities is that some ideal (as in imagined) learner has to be factored into the planning. Typically this will be the learner who is considered able to the activity, thus setting up exclusion even before the lesson is enacted and if activities are 'designed with only the included participants in mind, the excluded seem not to fit in because of something in their own nature' (Minow 1990, p. 21).

In contrast curriculum as inquiry opens up space for a variety of approaches and outcomes making it possible to teach to learners' strengths rather than try to deal with or compensate for perceived 'deficits', and comes about through building communities of inquiry.

From Collectives to Communities

One way of talking about a class is as a collection of, say, 25 individuals, each of whom has a specific level of mathematical understanding to be catered for, a view that can be perpetuated through the emphasis on planning for individual needs.

Another, complementary, way is to think of a class as a collective and treat learning as something that collectives—whole classes—engage in. Doing so changes expectations from the sum of the needs of the individuals to the potential of the collective and the recognition, in the words of Surowiecki (2005), that 'the collective can be smarter than the collection'. Shifting learning from the individual to the collective challenges the dominant discourse of 'individual needs' and 'personalized learning' and the associated 'good' practices such as grouping by 'ability', but this shift need not be at the cost to the individual. Raising expectations for the collective leads to individual higher attainments in ways that the reverse, bottom up, approach from the individual to the collective may not (Sapon-Shevin 1999). Evidence that attending to the collective, treating classrooms as communities, means welcoming diverse contributions, playing down differences and promoting inclusion (Smith and Barr, op cit., p. 407).

Moving from a view of teaching as having to meet the individual needs of a diverse collection of learners to looking at the power of and setting directions for the collective means having a clear sense of the sort of collective, the type of community, that is being fostered. Just as there are different interpretations of curriculum, so too there are different models of community each nuanced with regard to the relationship between community and learning. The work of Watkins (2005) is helpful here.

Watkins characterises community in three ways, three 'levels' that envelope each other. The first level is for schools and classrooms to operate as communities in the sense of having supportive and respectful relationships between all the members—staff, students and parents. It is beyond the scope of this chapter to look at the literature on schools as communities—my focus is on classrooms—but I note that the findings on schools needing to be communities is in line with complexity theory and a view of schools as systems that are emergent, self-organising and fractal in nature with the parts mirroring the whole and vice-versa (Davis and Sumara 2006). In other words, there is a complementary relationship between schools as communities and classrooms as communities, each influencing the other. Although it may be easier to establish classrooms as communities if schools are functioning as communities, that does not mean classrooms cannot function as communities in the absence of such a spirit at the school level and that, indeed, creating classroom communities may be the key to creating school communities.

While establishing classroom community may be necessary to support inclusive mathematics teaching, it is not, however, sufficient to ensure this. Classrooms as communities may focus primarily on establishing good relationships without necessarily questioning views of learning or the norms of pedagogy and as Watkins points out it is possible for classrooms to be communities and yet the teaching still to be teacher centred and not necessarily inclusive.

Watkin's next 'level' of community then is 'classrooms as communities of learners'. This, he argues, is distinguished from classrooms as a community through three aspects:

- A focus on intentional learning that engages with the discipline;
- Students learning from each other and helping each other to learn, and,
- Students being motivated towards learning for its own sake, making choices and being responsible (adapted from Watkins 2005).

Communities of learners, Watkins argues, characterises classrooms where pedagogy is predicated on a view of learners as individual sense-makers (that is, a view of learning based on constructivists principles) and does lead to better individual learning outcomes. Thus, Watkins' model of community of learners fits with Smith and Barr's model of curriculum as activity. Beyond establishing good relationships inclusion is encouraged because 'all students are present, engaged in culturally valued learning, participating and feeling they belong in their classroom community of learners' (Carrington and Macarthur, p. 270).

But Watkins argues that communities of learners can be developed further towards 'learning communities'. Communities of learners still primarily emphasise the individual, albeit individuals who support each other. Classrooms as learning communities mark the shift to focusing on the collective, that the learning unit is the whole. Drawing again on Watkins, the distinct characteristics here include:

- Disciplined discourse,
- Rich, co-constructive conceptions of learning,
- Developing shared metacognition about the process of learning, and
- Shared responsibility for and control of knowledge (adapted from Watkins 2005).

In learning communities, individuals are members of a knowledge building community and the increase in the collective knowledge generated is what promotes the growth in individual knowledge (Bereiter and Scardamalia 1996).

Thus, the emphasis shifts further, from expectations of what individuals might learn to collective development of mathematical ideas, based in curriculum as inquiry. Classrooms as learning communities promote inclusivity and diversity because of the increased range of roles that learners can play. Classrooms as learning communities are models of democratic education.

Developing learning communities means providing opportunities for engaging with mathematics that are authentic, start from mathematical activity and processes and are structured to support mathematics to emerge. Davis and Simmt (2003) suggest that there are five conditions necessary for emergence of this sort:

- Diversity,
- Redundancy,
- Enabling constraints,
- Neighbour interactions, and
- Distributed control.

Diversity means a need for variation amongst the participants and it is necessary to the possibilities for novel responses. If classes or groups are too homogeneous, then the chances for new mathematical ideas to emerge are reduced.

Redundancy is the other side of the diversity coin: members of a community have to have sufficient common ground, rules and assumptions to be able to work together. Davis and Simmt suggest that for emergence of new ideas, redundancy is helpfully thought about in terms of proscription—what we do not do round here—rather than prescription—we only do things this way.

Enabling constraints may sound like an oxymoron but these provide focus to activity while still allowing for diversity. For example, requiring students to work in pairs on a problem with only one piece of paper and one pen between them imposes a constraint that enables joint understanding to emerge that are greater than either individual started with.

Neighbour interactions means more than students working together. It means the sharing of ideas, hunches, questions and records of solutions. It means dialogue about the outcomes of problem solving, the reasons for these and what mathematics the community values.

Distributed control is probably the one area that most challenges current pedagogies as it means teachers having to relinquish being at the centre of mathematics lessons.

Building Learning Communities

Establishing classrooms as learning communities is a continuous and emergent process and never complete: the processes of community building have to be engaged in across the whole school year and between different years of schooling. When continuously attended to community building becomes

> the foundation that supports cooperative learning, differentiated instruction, and the formation of positive classroom relationships and talk. To ensure successful implementation, teachers intentionally create classrooms that engender a sense of safety and belonging, value diversity, share responsibility for the community, and an overall atmosphere of support and caring. (Reid and Valle 2004, p. 475)

Community building is an important aspect of promoting learner engagement, which can be examined through D'Amato's (1992) distinction between structural and situational rationales for learning. Structural rationales draw on the extrinsic values attached to learning (mathematics), such as high scores on tests easing entry into further study, or (as is the case in some high-attaining nations) as a way of bringing honour to one's family (Askew et al. 2010). Situational significance arises through local value 'in which students view their engagement in classroom activities as a means of maintaining valued relationships with peers and of gaining access to experiences of mathematics and accomplishment' (Cobb and Hodge 2007, p. 165).

De Abreu and Cline (2007) discuss this issue in terms of social valorisation, a dynamic concept describing dominant views and the value attached to learning mathematics not only in society at large (the structural) but also within local communities (the situational). The values that come to be attached to learning mathematics, De Abreu and Cline argue is a 'relational' construction and similar practices may either be valued or de-valued depending on the network of relations in which they are located.

> We took the view that the mathematical practices as products of the cultural heritage do not exist in a social vacuum, but are owned by social groups, whichever position in a social

> order. This enabled us to theorise the process of mathematical learning in a social–psychological terms. (De Abreu and Cline 2007, p. 122)

They draw on Tajfel's (1978) work that views the constructs of social categorisation and social comparison both as complimentary to social identity. Social categorisation and social comparison are the frameworks available to individuals from the overall social world while social identity is how individuals then place themselves within these frameworks. De Abreu and Cline see this as a valuable adjunct to Vygotskian theory in that

> while Vygotsky's theory focused on how the interpersonal knowledge became intrapersonal knowledge, Tajfel's theory examines how individuals come to adopt social identities from the social groups available in the broader social structure. …These two theories come together when one adopts the view that mathematical learning and thinking develop through forms of participation in the practice of specific communities, each of which has a position in the structure of the society that is often well understood by all. (De Abreu and Cline, op. cit., pp. 122–123)

The study of stereotypical threat would be a case of Tajfel's theory playing out. All of us are subject to being stereotyped, not in the pejorative sense that is often implied by the term, but in the sense of moving outside our usual communities of practice: a professor in a suit wandering into a leather biker's bar is going to be stereotyped. The flip side of this is that we do identify with being a member of different groups and research evidence is showing how being tuned in to being members of particular groups can either depress attainment (stereotype threat) or, usually less dramatically, raise it (stereotypical lift). For example, female Asian high-school students performed differentially on a mathematics test when randomly assigned to groups being 'tuned into' either their 'female' or 'Asian' identities (or to a control group). The 'female' group performed worse on a mathematics test in comparison to the control group—stereotypical threat—while the 'Asian' group scored higher—stereotypical lift (Spencer et al. 1999).

Thus, while the bringing into awareness of social identity—Asian, female, etc.—had a real impact on scores on mathematics tests, this stereotypical threat research shows the complexity of such self-identification (which I consider to be a preferable term to social identity) as we all have multiple social identities. This points to the importance of groups that learners identify with and, again, the importance of community. A sense of 'we are a community that does well' needs to be fostered, not 'we are a collection of learners where some do better than others'. But equally, there is little point in having high expectations for a collective, if the individual learner has little desire to be identified as part of that collective.

When sound situational rationales are thus established—and community building is a key element of this—the community members (teachers and students) adopt an attitude of 'this is just how we are around here', promoting a collective view of everyone as learners of mathematics. Gutiérrez, for example, reports on a study of senior high-school students on the calculus track—students who, stereotypically, would not have been in that group. Because of the strong sense of community created the students played down the 'high' status of being in that class and 'the fact that they saw a kind of 'normality' in their participation signals to me that they had incorporated aspects of this learning environment into their identities' (Gutierrez 2007, p. 44).

Relational and Attentive Listening: The Key to Learning Communities?

Curriculum as inquiry and building learning communities requires a shift from the (still) typical initiation–response–feedback (IRF) (Edwards and Westgate 1987) pattern of many classroom interactions towards dialogue between all participants (learners and teachers). Much is written on the importance of talk in mathematics classrooms and here I only briefly highlight aspects of talk that are pertinent to enacting curriculum as inquiry. I use the bulk of space to examine an aspect of communication that receives rather less attention than talk—listening. Listening is central both to generating knowledge and promoting inclusion.

Bohm (2004) makes the distinction between discussion and dialogue—the former being associated with holding to a position that one is trying to convince the other of adopting. Dialogue, in contrast, is more characterised by exchange of ideas, which requires holding lightly to one's position.

A shift towards dialogue brings advantages that extend beyond simply improving standards in mathematics. As Robin Alexander points out, dialogue is at the heart of developing caring learners:

> Dialogue requires willingness and skill to engage with minds, ideas and ways of thinking other than our own; it involves the ability to question, listen, reflect, reason, explain, speculate and explore ideas; to analyse problems, frame hypotheses, and develop solutions; … Dialogue within the classroom lays the foundations not just of successful learning, but also of social cohesion, active citizenship and the good society. (Alexander 2006)

Curriculum as inquiry and creating learning communities means attending to relations over and above simply fostering good classroom atmospheres. Listening is at the heart of this attending. Rinaldi (2001), for example, characterises listening as a relational process based around being 'orientative, curious and responsive as opposed to pre-determined, disinterested and pre-scriptive' (p. 3). In a similar vein, Veck (2009) distinguishes between the 'disciplinary gaze' characterised by 'the end of listening' and the 'attentive gaze' that seeks understanding through listening. Writers like Rinaldi and Veck come to these positions from an interest in pedagogies that promote democratic classrooms but the mathematics education research evidence shows that inclusive and democratic pedagogies also lead to better mathematical learning (see for example, Boaler 1997; Zhang et al. 2011).

Yet relational and attentive listening seem difficult for teachers to enact, as colleagues and I witnessed in a project examining teaching mental calculation strategies. Every lesson we observed was based around teaching towards a predetermined strategy, for example, using compensation for addition by rounding a number to a multiple of ten, adding that and adjusting the answer (e.g. $37+38=37+40-2$). As the lessons unfolded strategies offered by learners that did not fit with the prestated learning outcomes were gently dismissed or reframed to try and fit with expectations. For instance, in a lesson on compensation, one girl's strategy was based on near doubles—she knew that double 38 was 76 so $37+38$ was 75. The teacher tried, not very successfully, to fit the girl's explanation with her model of compensation that was the focus of the lesson (Askew et al. 2003).

I am not advocating, however, that teaching simply involves throwing a bunch of examples together and then 'winging it' by seeing what students come up with. Preparation for listening involves careful choice of examples. In a lesson on compensation $37+38$ may be less likely to provoke a near-doubles strategy than choosing to work with, say, $37+48$. On the other hand, $37+38$ might be chosen precisely because it may well provoke a variety of solution strategies. It is not that one example is better than the other but that examples have to be chosen in anticipation of strategies likely to emerge.

The demands that relational and attentive listening place on teachers' mathematics subject knowledge for teaching (SKT) may be one reason why such listening is hard to enact as it requires making sense, in the moment, of learners' ideas. But this demand on teacher knowledge can be eased. Techniques such as 'revoicing'—getting other learners to explain in their own words what they think a peer said—provides time and space for everyone in the class—the teacher included—to process and make sense of what is being said.

Breaking set with the idea of the lesson as the primary 'unit of learning' can also help as student work can then be considered between lessons. New technologies open up new opportunities for this. Students working on tablets can record short 'movies' that capture the construction of images and solutions along with a 'voice-over', allowing for listening outside actual lesson time.

Relational and attentive listening means treating learners as already competent in communicating mathematics. Moschkovich summarises a competence view of bilingual learners, but her points extend to diverse learners generally. She argues that communicative competence encompasses three things:

- Regarding communication as more than just spoken language (to includes things such as gesture).
- Seeing meanings as 'multiple, changing, situated and sociocultural'.
- Acknowledging that diverse learners may be different from each other but not defining such differences in terms of deficiencies (Adapted from Moschkovich 2007).

Moschkovich argues that growth in mathematical talk occurs through learners taking part in mathematical practices in whatever way they can, thus reframing diversity away from being an obstacle to classroom talk being enriched through the diversity of learners' contributions, even if these are mathematically limited.

This is in line with work a colleague, Penny Latham, and I were involved in a low-achieving school in London where opportunities that the teachers provided for learners to talk about mathematics were limited because of a shared perception amongst the staff that the children's lack of experience meant they were not capable of engaging in mathematical dialogue. The primary focus of our intervention was to encourage learners to talk about mathematics in whatever ways they could and thus build competency over time. Although ours was not the only intervention, the school has changed to now being regarded as one of the most highly attaining schools in the country (Askew and Latham 2005).

A challenge then is for teachers to set aside expectations of correct mathematical talk and work with the communicative resources that students do bring to school, including gestures and social resources. Yet while accepting that there is no one singular mathematical discourse we must also recognise that 'in general common abstract in, generalising, social certainty, and being precise, explicit, brief, and logical hardly are highly valued activities across different mathematical communities' (Moschkovich, op. cit., p. 95).

Relational and attentive listening also poses challenges to learners. They need to listen to each other's solutions, not just out of politeness, but to think about the connections to their solutions and to help each other refine methods and explanations.

The sharing of methods can, legitimately, be followed up with asking learners to try out a particular approach and see if other problems could be done in that way. The evidence from psychology is that this leads to better learning (Langer 1997). Being, in Langer's term, mindful of the fact that there are choices—that, say, a calculation could be done in several ways—results in deeper and more flexible learning.

Working towards everyone listening relationally and attentively not only has a focus on the collective, not only can raise standards of attainment, but is at the heart of dialogue and the welcoming of diversity. Without these there the dangers that Paolo Freire warns of may emerge:

'Leaders who do not act dialogically, but insist on imposing their decisions, do not organize the people—they manipulate them. They do not liberate, nor are they liberated: they oppress' (Freire 2000).

References

Alexander, R. (2006). *Towards dialogic teaching: Rethinking classroom talk*. Cambridge: Dialogos.

Askew, M., & Latham. P. (2005). *Learning from practice: Teachers and researchers in joint enquiry*. Proceedings of the fifteenth international commission on mathematical instruction (ICMI) study: The professional education and development of teachers of mathematics. Brazil.

Askew, M., Bibby, T., Brown, M., & Hodgen, J. (2003). Mental calculation: Interpretations and implementation. In N. A. Pateman, B. J. Dougherty, & J. T. Zilliox (Eds.), *Proceedings of the 2003 joint meeting of the international group for the psychology of mathematics education (PME) and PME North America* (Vol. 1, pp. 1–202). Honolulu: CRDG, College of Education, University of Hawaii.

Askew, M., Hodgen, J., Hossain, S., & Bretscher, N. (2010). *Values and variables: Mathematics education in high performing countries*. London: Nuffield Foundation.

Ball, D. L., Hill, H. C., & Bass, H. (2005). Knowing mathematics for teaching: Who knows mathematics well enough to teach third grade, and how can we decide? *American Educator, Fall 2005,* 14–17, 20–22, 43–46.
Bereiter, C., & Scardamalia, M. (1996). Rethinking learning. In D. R. Olson & N. Torrance (Eds.), *The Handbook of education and human development: New models of learning, teaching and schooling* (pp. 485–513). Cambridge: Blackwell.
Boaler, J. (1997). *Experiencing school mathematics: Teaching styles, sex and setting.* Buckingham: Open University Press.
Bohm, D. (2004). *On dialogue.* New York: Routledge Classics.
Carrington, S., & MacArthur, J. (2012). *Teaching in inclusive school communities.* Milton: Wiley.
Cobb, P., & Hodge, L. L. (2007). Culture, identity, and equity in the mathematics classroom. In N. S. Nasir & P. Cobb (Eds.), *Improving access to mathematics: Diversity and equity in the classroom* (pp. 159–171). New York: Teachers College Press.
Cockcroft, W. H. (1982). *Mathematics counts: Report of the Committee of Inquiry into the teaching of mathematics in schools.* London: Her Majesty's Stationery Office.
D'Amato, J. (1992). Resistance and compliance in minority classrooms. In E. Jacobs & C. Jordan (Eds.), *Minority education: Anthropological perspectives* (pp. 181–208). Norwood: Ablex Publishing.
Davis, B., & Simmt, E. (2003). Understanding learning systems: Mathematics education and complexity science. *Journal for Research in Mathematics Education, 34,* 137–167.
Davis, B., & Sumara, D. (2006). *Complexity and education: Inquiries into learning, teaching and research.* Mahwah: Lawrence Erlbaum.
De Abreu, G., & Cline, T. (2007). Social valorization of mathematical practices: The implications for learners in multicultural schools. In N. S. Nasir & P. Cobb (Eds.), *Improving access to mathematics: Diversity and equity in the classroom* (pp. 118–131). New York: Teachers College Press.
Dweck, C. S. (2000). *Self-theories: Their role in motivation, personality, and development.* Philadelphia: Psychology Press.
Edwards, A. D., & Westgate, D. P. G. (1987). *Investigating classroom talk.* London: Falmer.
Freire, P. (2000). *Pedagogy of the oppressed.* London: Bloomsbury Publishing.
Gutiérrez, R. (2007). (Re)defining equity: The importance of a critical perspective. In N. S. Nasir & P. Cobb (Eds.), *Improving access to mathematics: Diversity and equity in the classroom* (pp. 37–50). New York: Teachers College Press.
Kavale, K. A., & Forness, S. R. (1998). The politics of learning disabilities. *Learning Disability Quarterly, 21*(4), 245–273. doi:10.2307/1511172.
Ladson-Billings, G. (1997). It doesn't add up: African American students' mathematics achievement. *Journal for Research in Mathematics Education, 28*(6), 697–708. doi:10.2307/749638.
Langer, E. J. (1997). *The power of mindful learning.* Cambridge: Da Capo Press.
Minow, M. (1990). *Making all the difference: Inclusion, exclusion, and American law.* Ithaca: Cornell University Press.
Moschkovich, J. (2007). Bilingual mathematics learners: How views of language, bilingual learners, and mathematical communication affect instruction. In N. S. Nasir & P. Cobb (Eds.), *Improving access to mathematics: Diversity and equity in the classroom* (pp. 89–104). New York: Teachers College Press.
Oakes, J. (2005). *Keeping track: How schools structure inequality* (2nd ed.). New Haven: Yale University Press.
Reid, D. K., & Valle, J. W. (2004). The discursive practice of learning disability: Implications for instruction and parent-school relations. *Journal of Learning Disabilities, 37*(6), 466–481.
Rinaldi, C. (2001). A pedagogy of listening. *Children in Europe, 1*(1), 2–5.
Robitaille, D., & Dirks, M. (1982). Models for the mathematics curriculum. *For the Learning of Mathematics, 2,* 3–21.
Sapon-Shevin, M. (1999). *Because we can change the world: A practical guide to building cooperative, inclusive classroom communities.* Boston: Allyn and Bacon.

Small, M. (2012). *Good questions: Great ways to differentiate mathematics instruction. 2nd edition.* New York: Teachers College Press.
Smith, R., & Barr, S. (2008). Towards educational inclusion in a contested society: From critical analysis to creative action. *International Journal of Inclusive Education, 12*(4), 401–422.
Spencer, S. J., Steele, C. M., & Quinn, D. M. (1999). Stereotype threat and women's math performance. *Journal of Experimental Social Psychology, 35*(1), 4–28. doi:10.1006/jesp. 1998.1373.
Surowiecki, J. (2005). *The wisdom of crowds: Why the many are smarter than the few*. London: Abacus.
Tajfel, H. (1978). Social categorization, social identity and social comparison. *Differentiation between social groups: Studies in the social psychology of intergroup relations* (pp. 61–76). London: Academic.
Task Group on Assessment and Testing (TGAT). (1988). A report. London: Department of Education and Science and the Welsh Office.
Varenne, H., & McDermott, R. (1999). *Successful failure: The school America builds*. Boulder: Westview Press.
Veck, W. (2009). Listening to include. *International Journal of Inclusive Education, 13*(2), 141–155.
Von Glasersfeld, E. (1990). An exposition of constructivism: why some like it radical. In R. B. Davis, C. A. Maher, & N. Noddings (Eds.), *Constructivist views on the teaching and learning of mathematics*. Reston: NCTM.
Watkins, C. (2005). Classrooms as learning communities: A review of research. *London Review of Education, 3*(1), 47–64. doi:10.1080/14748460500036276.
Wiliam, D., & Bartholomew, H. (2004). It's not which school but which set you're in that matters: The influence of ability grouping practices on student progress in mathematics. *British Educational Research Journal, 30*(2), 279–293.
Zhang, J., Hong, H. Y., Scardamalia, M., Teo, C. L., & Morley, E. A. (2011). Sustaining knowledge building as a principle-based innovation at an elementary school. *The Journal of the learning sciences, 20*(2), 262–307.

Chapter 9
Ethics and the Challenges for Inclusive Mathematics Teaching

Helen Forgasz, Jennifer Bleazby and Carly Sawatzki

Introduction

The recently developed Australian curriculum encompasses an innovative and contemporary set of goals and expectations. Ethical understanding is included as one of the seven general capabilities that learners will develop, along with literacy, numeracy, information and communication technology, critical and creative thinking, personal and social capability, and intercultural understanding. These general capabilities are to be incorporated inclusively across all discipline domains. The general capabilities encompass the knowledge, skills, behaviours, and dispositions that, together with curriculum content in each learning area and cross-curriculum priorities, must be taught by all teachers. It is claimed that these general capabilities will assist students to live and work successfully in the twenty-first century. That all students should develop these capabilities over the course of their schooling is consistent with the Melbourne Declaration on Education Goals for Young Australians (MCEETYA 2008) in which it was advocated that all young people should become "confident and creative individuals and active and informed citizens" (np). The general capabilities also reflect the emphasis placed on preparing students to "work for the common good, in particular sustaining and improving natural and social environments" and be "responsible global and local citizens" (MCEETYA 2008, np).

Ethical understanding, a welcome inclusion as a general capability in the Australian Curriculum, has been defined by the Australian Curriculum, Assessment and Reporting Authority (ACARA) (2012a) as follows:

H. Forgasz (✉) · J. Bleazby · C. Sawatzki
Faculty of Education, Monash University, Clayton, Australia
e-mail: helen.forgasz@monash.edu

J. Bleazby
e-mail: jennifer.bleazby@monash.edu

C. Sawatzki
e-mail: carly.sawatzki@monash.edu

A. Bishop et al. (eds.), *Diversity in Mathematics Education*,
Mathematics Education Library, DOI 10.1007/978-3-319-05978-5_9,

> Students develop ethical understanding as they learn to recognise and understand matters of ethical concerns, make reasoned judgments and, in so doing, develop a personal ethical framework. This includes understanding the role of ethical principles and values in human life, acting with integrity and regard for the rights of others, and having a desire to work for the common good. (p. 21)

The ethical understanding learning continuum includes three interrelated organising elements:

- Understanding ethical concepts and issues;
- Reflecting on personal ethics in experiences and decision-making;
- Exploring values, rights, and ethical principles.

In the Australian Curriculum, Mathematics (F-10), the following explication is found within the section on general capabilities under the heading *Ethical behaviour:*

> Students develop the capability to behave ethically as they identify and investigate the nature of ethical concepts, values, character traits and principles, and understand how reasoning can assist ethical judgment. Ethical behaviour involves students in building a strong personal and socially oriented ethical outlook that helps them to manage context, conflict and uncertainty, and to develop an awareness of the influence that their values and behaviour have on others.
>
> There are opportunities in the Mathematics curriculum to explore, develop and apply ethical behaviour in a range of contexts, for example through analysing data and statistics; seeking intentional and accidental distortions; finding inappropriate comparisons and misleading scales when exploring the importance of fair comparison; and interrogating financial claims and sources. (ACARA 2011, p. 11)

We believe that the examples noted as opportunities for exploring ethical issues within the mathematics curriculum are limited and reflect a narrow conception of what is possible in promoting students' evolving ethical and moral development. The mathematics curriculum also advocates exposing students to a range of diverse real world contexts that are familiar to them. There are few situations in real life that are devoid of ethical and moral dilemmas and concerns. We are concerned, however, that mathematics teachers generally steer clear of the controversies inherent in many of the contexts they select for students to engage in problem solving, believing that this is the purview of other disciplines in the school curriculum, or are better dealt with at home. With respect to ethical considerations, this level of avoidance may be exacerbated.

We recognise that ethics, values, morals, and social justice considerations are not absolute or consistent across national, cultural, or religious boundaries. In this chapter, we do not propound our personal ethical positions, nor do we advocate in favour of one perspective over another. Some teachers are likely to have defensible reasons for not wanting to confront or discuss certain contentious ethical issues because of who they are or where they work. This is taken as a given, and it is not advocated that every potential opportunity to explore ethical considerations within mathematical contexts must be pursued. In our discussion in this chapter, we highlight examples in which the opportunities to promote all students' ethical development exist. In Chapter 10 of this volume, Seah and Andersson explore a related notion: how mathematics teachers' personal values in the context of mathematics and its teaching can be enacted in the mathematics classroom.

We commence this chapter with each author discussing the impetus for writing about ethics and mathematics learning. We then examine the philosophical and psychological bases of children's moral development in the educational context, and review earlier thinking on ethical, moral, and social justice dimensions of mathematics learning. We explore a variety of contexts within particular mathematics content domains that are commonly used when teaching, and which text book writers employ, and expose the ethical implications of these contexts. We demonstrate, with examples, how the contextual setting of the mathematics problems that students grapple to solve can be used in the mathematics classroom to address inclusively the *ethical understanding* capability of the Australian mathematics curriculum. We conclude with a discussion of implications for classroom practice and professional preparation, and suggest future research directions.

Impetus for Writing about Ethics and Mathematics Learning

Helen's Story

Not so long ago I was at a professional development session for mathematics teachers. The principal of the school at which the session was held welcomed participants. As a former head of mathematics in a high school, the principal reflected on the importance of mathematics for the future of all children and reminded those present how privileged they were to be teaching this critically important discipline. He spoke of the relevance of mathematics and its power to model reality. The following question was posed as an example, "Imagine you are the general of three army divisions. The first is winning handsomely, the second is holding its ground, and the third is suffering huge losses. You have sufficient reserve troops to send in as support for one division only. Where would you deploy these troops?" The answer, the audience was told, was simple and based on mathematical modelling, "To the winning division, naturally". The principal gave a second example. "Imagine you are charged with placing landmines to produce maximum kill. How would you arrange them?" Again, he claimed, mathematical modelling would enable this decision to be made very easily.

My equilibrium was disturbed. I felt an uncomfortable disquiet as I left the meeting. The two examples used to exemplify the power of mathematics were in military contexts. What messages could be inferred from them? From the first, one could infer that mathematics was contributing to the military objective of winning a battle and that it was acceptable that the lives of at least one group of soldiers were expendable. A possible interpretation of the second example was that it is acceptable to use mathematics to promote killing, in this case to calculate a maximum kill rate through landmine placement. Had the principal reflected on how "realistic" and relevant these examples were to the lives of those in the room and for most school students? And what about the covert messages potentially conveyed? On the surface, the principal offering these examples did not appear concerned that what he had said might be considered offensive by those present, or that the examples might be considered unacceptable in a school context. He seemed oblivious that anything might conceivably be wrong with what he had put forward.

As the man spoke, there were astonished expressions on the faces of several members of the audience. No-one, including me, was prepared to get up and say anything. Polite acceptance of what had been said left this man totally unaware of the impact that his words may have had. The experience left an indelible mark on me. Driving home, I reflected on the examples that I may have used when teaching mathematics in schools when I was a high school teacher, or the examples I may have suggested to the preservice mathematics teacher education students I had taught. What messages might the students have picked up? If I was guilty of adopting similar "real world" mathematical examples with unsavoury implications, what about other mathematics educators and teachers? I was distressed that some of my students may have picked up on the covert messages conveyed but were uncertain or loath to voice their discomfort or concerns.

Reflecting more deeply about the issue, I realised that it was broader than first imagined. I remembered reading, and later writing a journal editorial (Forgasz 2005), about the "Lawrence Summers incident". Here was another educational leader (President of Harvard University) who appeared not to have thought through the consequences of what he was about to say. In a talk at an academic conference, he implied that there were innate (genetic) differences between men and women which may partially explain why there were fewer female than male science academics at Harvard and why women were not as successful as men in academic science and mathematics careers. Interpretations of exactly what he had said differed. Some walked out of the talk disgusted, while others were not at all offended (see Bombardieri 2005). Because people spoke out, the ensuing controversy and uproar resulted in Summers resigning his post at Harvard.

Jen's Story

In 2008, while the events of the Global Financial Crisis continued to unfold, I was teaching Year 12 philosophy at a government school in Melbourne. My philosophy students had been studying a unit called "The Good Life" which involved a close study of texts from Plato and Aristotle. A key theme throughout this unit was whether a hedonistic life—a life devoted to satisfying pleasures—could be a good life. Plato and Aristotle provided many reasons to support the view that a good life was actually one of temperance.

In their writing and classroom discussions, many students used the Global Financial Crisis to support a modern form of Platonism, arguing that living beyond one's means, greed, materialism, and consumerism could lead to considerable unhappiness for oneself and for society at large. The course content had provided my students with an invaluable opportunity to study the ethical, political and social implications of this historical event as it unfolded. However, this made me wonder if students in other subjects, especially mathematics, business, and economics, were doing the same.

An opportunity arose for me to do some team teaching, resulting in me taking a Year 11 mathematics class. They were completing a unit on financial mathematics. The students were given an overview of the causes and implications of the Global Financial Crisis, before looking at the related issues of consumerism and high personal debt levels amongst young people in Australia. Students were then given an opportunity to discuss some of the ethical and social issues associated with relevant financial systems and practices. The students seemed genuinely interested and concerned, and appreciated the opportunity to learn about and discuss these significant issues and events. However, this was a one-off class, which seemed problematic, especially given that most of these mathematics students were not enrolled in philosophy or other humanities subjects where they would be able to consider such ethical issues. Surely these students, many of whom wanted to go into fields like business, accounting, or science, needed a thorough, critical understanding of the ethical issues related to the fields they wished to work in and study.

Carly's Story

My interest in integrated approaches for teaching Mathematics together with ethics stems from my PhD study in which I examine the role of attitudes and values in financial literacy teaching and learning. Financial literacy teaching and learning is complex for a range of reasons, not least of which being that it is socially constructed and situated. This means that students bring to the classroom preconceived ideas about money and how it might be earned, spent, saved, and shared based on what they have observed, heard about, and experienced at home and in the community. Parents are significant agents of economic, consumer, and financial socialisation in childhood and adolescence; their attitudes and values regarding money, together with how much and how well they educate their children about money can contribute significantly to their children's financial motivations, financial literacy learning, and financial behaviour (Danes 1994; Lewis and Scott 2003; Webley and Nyhus 2006; Shim et al. 2009, 2010).

Where money is concerned, profit maximisation and self-interest tend to be priorities. Whether a financial decision is being made by an individual, household, business, or government, it tends to involve the question, "How will I be better off?" This way of thinking is counterproductive to a sustainable future to the extent that it often results in choices that have a negative impact on local and/or global communities, and/or the environment. In the twenty-first century, it is important to teach students that being "better off" is not simply a matter of dollars and cents—there are significant ethical, social, and environmental considerations that should be carefully weighed up before deciding how to earn, spend, save, and share money are made. This suggests the selection of pedagogies that encourage a sense of social and environmental responsibility, and that foster ethical maturity.

While financial literacy education at school tends to focus on developing students' financial knowledge and skills, the behavioural economics research literature builds a compelling case that financial behaviour may depend more on intrinsic psychological attributes together with attitudes and values learned at home, than teachable knowledge and skills (Homer and Kahle 1988; de Meza et al. 2008). What then, are the implications for educators? My work involves exploring the opportunities that exist to weave together the teaching of numeracy and ethics (values) in authentic and inclusive ways so as to enhance financial literacy teaching and learning. The issue is not whether attitudes and values learned at home are "right" or "wrong", but that the educational process could be improved by examining them, and promoting thoughtful consideration about adopting them (Gray et al. 1994).

Philosophical and Psychological Antecedents of Ethical Development

Expanding on Piaget's developmental theory, Lawrence Kohlberg set out to describe the distinct types of moral reasoning that characterise different stages of human development. Kohlberg was concerned with the justifications that people give for their moral beliefs, more so than with the content of those beliefs or with moral behaviours. This is because an individual's moral beliefs and behaviours are not necessarily indicative of their level of moral maturity (Duska and Whelan 1977). Two individuals may form the same moral judgment (e.g., that stealing their favourite library book would be wrong) and may even behave in the same way (e.g., begrudgingly returning the book to the library) but they may have very different levels of moral maturity. The first individual may have employed higher order thinking processes to formulate a judgment about what action to take, while the second person may have unthinkingly followed externally set rules or social conventions. In this situation, the first person is the more morally mature because more sophisticated and autonomous moral inquiry skills have been applied. Unlike person two, person one is not entirely dependent on external deterrents or guides to formulate moral judgments. Thus, an assessment of one's level of moral maturity necessarily involves observing the reasoning process used to formulate moral judgments.

For this reason, Kohlberg used moral dilemmas in his empirical research. The most well known is the Heinz dilemma. A man named Heinz breaks into a pharmacy and steals the only drug that will save the life of his dying wife, who has a rare form of cancer. The pharmacist was charging ten times what the drug costs to make and refused to give Heinz a discount despite knowing that Heinz's wife was dying and that Heinz did not have enough money to pay the full price (Duska and Whelan 1977). After being presented with such a moral dilemma, research participants were questioned about the morality of Heinz's action and asked to explain the reasons that had led them to this moral judgment. This research enabled Kohlberg to formulate six sequential stages of moral development. According to Kohlberg, the earlier stages of moral development are characterised by thinking that is more egocentric and concerned with obeying authority so as to avoid punishment or disapproval from others. As individuals develop, their thinking gradually becomes less egocentric and more autonomous, objective, and sophisticated. For example, the sixth and final stage is characterised by a focus on logic and universal ethical principles and human rights.[1] Kohlberg maintained that all individuals move through the stages in the same order. However, not all individuals progress through the stages at the same rate, and not all reach the highest stages of moral development.

Kohlberg's six stages, and the very notion of universal developmental stages, are contentious. For example, Gilligan (1982) argued that Kohlberg's stages of moral development are not universal but reflective of a masculine moral perspective. Gilligan identified a feminine moral perspective that she referred to as the "care perspective". In contrast to the male "justice perspective", which emphasises emotional detachment, abstract reasoning, and universal principles, the care perspective focuses on the concrete particularities of moral situations, maintaining relationships, emotional responses, care, and empathy. Gilligan is sometimes misinterpreted to be claiming that the care perspective is superior to the justice perspective. However, she actually argues that effective moral problem solving incorporates valuable elements of both perspectives. That is, higher order moral thinking fully integrates the emotions, empathy, concrete thinking, logic, and general principles. Thus, Gilligan extended Kohlberg's theory, outlining a more inclusive, robust ideal of moral inquiry.

Despite their disagreements, both Kohlberg and Gilligan maintain that progression through the developmental stages, whatever those stages may be, is not merely internal to the individual, nor is it automatic. Piaget's, Kohlberg's, and Gilligan's developmental theories assume a Darwinian notion of growth, according to which development results from individuals' interactions with their environment, including their social and cultural environments. It means that particular conditions have to be present in the environment to stimulate growth. Interactionists deny the notion that if left to their own devices, individuals will naturally progress through the moral stages in accordance with their own internal mental capabilities. It is important to note that one may hold an interactionist theory of development and yet reject the notion that there are sequential stages of development. For example, Vygotsky (1978, 1986) and Dewey (1930, 2014 [1916]) also defended interactionist theories of development. Yet neither of them outlined Piagetian style developmental stages. What all interactionist theorists have in common is a belief that moral development is dependent on individuals interacting with their environments in particular ways. This means that even the mere presence of stimulating factors in one's environment is inadequate for growth. The individual actually needs to notice and interact with these conditions in the right way. The absence of such external conditions, or the

[1] For a more detailed account of each stage, see Kohlberg (1981) and Duska and Whelan (1977).

individual's failure to interact with them appropriately, will result in the retardation of moral development.

This has significant implications for education, implying that schools, curriculum designers, and teachers bear considerable responsibility for the moral development of students. Schools cannot assume that moral development is something that will occur naturally without teacher intervention. Moral development is dependent on teachers creating, within the classroom, the external conditions that stimulate moral development.

Furthermore, not just any type of curriculum and pedagogy will provide the conditions necessary to foster such an ideal of moral development. Transmissive pedagogies are inadequate because, as described, moral maturity is not simply a matter of having moral beliefs or even behaving in a moral way. Thus, simply telling students which actions are right and wrong, or asking them to memorise and obey particular moral rules are unlikely to promote moral development. Since moral maturity is reflected in the processes and strategies used to work through moral problems, teachers need to foster the capacities required for effective moral inquiry, such as moral reasoning and empathy. It is for this reason that Kohlberg (1987) drew upon the problem- and inquiry-based pedagogy promoted by Dewey in order to outline his ideal of moral education (see also Ísaksson 1979).[2]

In Dewey's laboratory school at the University of Chicago, students learnt important skills and knowledge through engaging in concrete, collaborative, problem-solving activities. For example, children were required to set up and run a shop, and this would stimulate the development of various literacy, numeracy, thinking, and social skills (Tanner 1997). The teacher's job was to create and maintain an environment that would stimulate the development of the desired skills. Thus, if the aim was to foster moral development, the teacher would need to identify the students' current levels of moral development, and design problem-based materials and tasks that prompt the students to critically reflect on any inadequacies with their current moral problem-solving strategies, as well as stimulate them to develop and use more sophisticated methods (Kohlberg 1987).

The moral problem-solving tasks also need to be carefully scaffolded by the teacher. It is not sufficient to simply present students with moral dilemmas and stand back and hope that they develop skills and knowledge needed to solve them. As explained, growth is dependent on individuals interacting with their environment in particular ways. Vygotsky's interactionist theory emphasises the fact that learners need to be scaffolded by a more knowledgeable other. Vygotsky argued that through engaging with others in activities that they could not perform by themselves, children can observe the skills, understandings, and behaviours of the more knowledgeable other and, through a process of internalisation, acquire the skills

[2] However, it should be noted that Kohlberg seemingly overlooked some fundamental differences between his ideas and those of Dewey. In particular, Kohlberg suggested that Dewey held a stage theory of development. However, Dewey would not have agreed with either Kohlberg's notion of universal, sequential stages of development or with the different types thinking that Kohlberg described as characterising each stage.

and knowledge needed to perform these tasks independently. Vygotsky famously referred to the space between what children can do with others and what they can do by themselves as the "zone of proximal development". This is the space in which children learn to act and think independently by being "scaffolded" by others (Vygotsky 1978). In relation to moral development, this means that the teacher should explain and model the more sophisticated moral inquiry skills that they want their students to develop. They should also ask questions, make suggestions, and intervene in the moral problem-solving tasks so as to prompt and guide students' development of the desired skills. For example, if a student responded to the Heinz dilemma by saying that what Heinz did was wrong because stealing is against the law, the teacher might prompt a more sophisticated answer by asking the student if there is such a thing as a bad law, or if it is ever acceptable to break laws. In collaborative problem-solving tasks, such scaffolding can also be provided by the students themselves.

So far we have argued that teachers bear considerable responsibility for their students' moral development. However, does this mean that all teachers bear this responsibility? In particular, are mathematics teachers responsible for fostering moral development? A common belief is that only teachers of philosophy, religion, personal development, or pastoral care are responsible for moral education. This belief is based on the false assumption that, unlike these other subjects, mathematics is a value neutral discipline where concepts like right, wrong, good, and bad are irrelevant. However, an inclusive approach means that every facet of human experience and all knowledge domains have ethical dimensions. In particular, mathematics is integral to many everyday situations and is implicated in many decision-making contexts. To the extent that mathematical knowledge and skills are often required to evaluate alternatives and make decisions that have social, economic, and/or environmental consequences, the mathematics classroom is an appropriate place to pose problems that embrace these ideas and which can promote and challenge students' ethical understandings.

A recent trend in mathematics teaching in Australia (and elsewhere) has been to contextualise the seemingly abstract content of mathematics precisely so that students learn how to apply mathematical thinking and concepts to "real life" situations, just as the children in Dewey's schools learnt numeracy through running a shop or building a garden shed. Contextualised, "real life" problems usually, if not always, contain ethical issues. For example, running a shop raises all sorts of ethical considerations, such as: the fair pricing of goods, just working conditions, ethics in advertising and labelling products, and the ethics of various economic practices (e.g., taxation). Presenting students with contextualised mathematical problems that raise obvious ethical issues, and then asking students to ignore the ethics and just focus on the mathematics, can send the rather dangerous message that, in mathematical contexts, ethical issues are secondary or can be disregarded altogether. Rather than feeling burdened by the responsibility to foster ethical inquiry, mathematics teachers should value such opportunities because they enable students to meaningfully connect the discipline of mathematics to other aspects of students' experiences. Dewey (2014 [1916]) argued that one of the benefits of contextualised "real life" problems is that they require students to draw on, and integrate, skills and knowledge from various academic disciplines, enabling them to understand the interrelatedness of the traditional knowledge domains.

The notion that moral education can be restricted to certain parts of the school curriculum also underestimates the amount of time and effort that needs to be devoted to moral development. This is particularly problematic if it is relegated to philosophy or religion classes or pastoral care sessions, subjects which are usually marginalized within the school curriculum, if they have any presence at all. Moral inquiry skills are as fundamental as thinking, literacy skills, and numeracy skills, in terms of students' capacities to function in society and live meaningful lives. This may be one underpinning driver for the identification in the Australian curriculum of the need for children to develop these skills as general, cross disciplinary, capabilities that must be fostered by all teachers in all subject areas.

Critical Mathematics Education: Democracy, Ethics and Morality, and Social Justice in the Mathematics Classroom

The emergence of the critical mathematics education movement in the early 2000s has presented opportunities for the discussion and inclusion of ethics in mathematics learning For some years, the mathematics education community has advocated that the mathematics taught in schools must have relevance to the lives of all students, and not simply focus on abstractions which privilege those who are most likely to persist with higher level mathematical studies leading to degrees and careers in the fields of mathematics, science, or finance; this is supported in the Australian curriculum—Mathematics (see Commonwealth of Australia 2009). Mathematics curricula in many countries have taken on board the call to provide "mathematics for all", although this direction has not been universally accepted and has some vociferous critics (e.g., see Schoenfeld 2003, for a history of the US Math Wars)

Mathematics focussing on everyday applications and meaning for all citizens has variously been called quantitative literacy (e.g., Steen 2001), mathematical literacy (e.g., de Lange 2003), or numeracy (AAMT 1998), the term adopted in Australia. Mathematical literacy is seen to extend beyond basic arithmetical competencies (a common notion of what it is to be numerate) to include algebraic, geometric, and statistical concepts. Mathematical literacy is different from the traditional notion of "applied mathematics", which is seen as complementing pure mathematics with high-powered, abstract, mathematical underpinnings. When implemented in the mathematics classroom, it is expected that teachers will identify relevant and familiar contextual settings for the examples and problems students are expected to solve; it is also assumed that teachers are willing and capable of doing so.

Educational researchers and commentators have written on critical dimensions of mathematical literacy (e.g., Frankenstein 2008; Gellert et al. 2001), mathematics and democracy (e.g., Skovsmose and Valero 2001; Thomas 2001), the social responsibility of mathematics (e.g., Ernest n.d.), and how these notions apply in the mathematics classroom (e.g., Mukhopadhyay and Greer 2001, 2002). Those writing about critical mathematics/literacy education emphasise the ethical, political, and social justice dimensions of mathematics and how it is used in society. In a provocative paper replete with examples drawn from the contemporary world at the time, Frankenstein (2008) advocates for the need to appreciate and unpack the quantitative forms in which data about the world are presented to the public: "[I]t is important to understand which aspect of quantitative evidence is mathematical fact and which is political, and therefore, subject to debate" (p. 268), and "to determine which quantitative measure gives the most accurate picture of a particular issue" (p. 268).

Children, we argue, need to grapple with the mathematical content and the intertwined moral, ethical, and social justice issues associated with the contextual setting of the problems they are asked to solve. Mathematics is found in a range of contexts in everyday life and is replete with critical dimensions. When mathematics is associated with competencies that teachers wish to develop in students, for example, readiness for the workplace, Skovsmose (n.d.; 2008) described this as *mathematics in action*. Applying mathematics to the "real" world, de Freitas (2008) argued, "requires recognizing both the messiness of life contexts and the limitations of the mathematical tools to adequately represent or model such messiness" (p. 87). Skovmose (2008) and de Freitas (2008) maintain that when *mathematics in action* is brought into the classroom, *ethical filtration* results, that is, the complexity of the inherent ethical, political, or social justice dimensions are reduced and simplified with the primary aim of reaching a numerical solution efficiently. Skovsmose (2008), however, provides an example of how an ethical perspective can be encompassed within a classroom mathematics problem which exemplifies mathematics in action, demonstrating the power of *mathematics in action* detached from *ethical filtration*.

The Australian curriculum now mandates that all teachers are expected to develop students' ethical understandings of the world. The question is how are the mathematics teachers to do this? What will assist and guide them in deciding which contexts are appropriate for doing so? To meet the ethical understanding goal of the Australian curriculum for mathematics, ethical filtration can no longer be universally applied. Mathematics teachers will need to reflect on appropriate mathematical contexts and problems, establish pedagogical processes, and implement strategies that encourage students to interrogate and consider core ethical issues and concepts including "justice, right and wrong, freedom, truth, identity, empathy, goodness and abuse" (ACARA 2012b, p. 78). ACARA (2011) maintains that the pedagogies should involve authentic, mathematically rich, cases which students can engage with by "giving reasons, being consistent, finding meanings and causes, and providing proof and evidence" (ACARA 2012b, p. 79). Appropriate resources will be required and teachers will have to be prepared to provoke and guide students in discussion of the relevant ethical issues encompassed in the problem contexts. In selecting appropriate contexts, sensitivity and respect for students' diverse backgrounds will also be needed. Professional development is likely to be in demand, and textbook writers and those producing resource materials will need to adapt.

Classroom Opportunities to Incorporate Ethical Understanding

Presenting students with routine mathematical problems that call for procedural approaches to arrive at a solution does little to facilitate the exploration of ethical issues or the learning of ethics/values, and students' moral development. The purpose of using socially situated investigative tasks that bring together numeracy and ethics/values is to facilitate learning through inquiry, and critical understandings of the complex factors associated with the context.

In what follows, we share examples from mathematics curricular sources which provide opportunities to promote students' moral development. We also discuss inappropriate contexts which have the potential to do the opposite. Our aims are to raise awareness of the potential to bring democracy and ethical considerations into mathematics classrooms and of the unintended learning that can result if teachers do not reflect critically on the selection of the contexts of the problems and exercises they require students to solve. Examples are drawn from mathematics textbooks and various online resources.

We begin with examples from the content domain of financial literacy, cited as an example for meeting the ethical understanding dimension in the Australian mathematics curriculum. Ethical considerations are central to consumer and financial understandings and decision-making. Then, examples drawn from other mathematical content areas including ratio and rates, projectile motion, data/statistics, and probability follow. We believe that the examples we present exemplify how ethical and moral dimensions of social justice and sustainability issues in our society can be brought into the mathematics classroom through a careful selection of resources. Clearly some topic areas and mathematical concepts lend themselves more readily to real life settings, and to ones that provide opportunities to develop students' ethical understandings.

Financial Literacy Examples

Consider the financial dilemma, reminiscent of Kohlberg's Heinz dilemma, aimed at primary school students shown in Box 1:

> Box 1
> Your mother asks you to buy a large carton of milk on your way home from school. The milk costs \$3, and you also buy an ice-cream for \$2. You pay with a \$10 note. The shop assistant is distracted by another customer and gives you \$10 too much change. What is the correct amount of change? What will you do about the error?

While the intended learning outcome for this problem is likely to be using mathematics to calculate change, there is clear educational merit in exploring the reasons children might give for returning or keeping the additional change. Will the students be more or less likely to return the money if the error involved less than $10? What if the error was $20? Do the children's views change if they know that the business owner requires the shop assistant to pay for the shortfall? Exploring the ethical and moral dimensions of this dilemma has the potential to promote consideration for and understanding of the impact that personal decisions can have on others.

The Oxfam Water for All website brings together quizzes, case studies, and mathematics exercises for teaching 9–13 year olds that water is an important, but limited, natural resource (Oxfam n.d.). The website activities also promote interdisciplinary learning with links across geography, English, mathematics, science, and citizenship. In the Water Maths: Global Statistics online quiz, students are asked to read, interpret, and match written and visual representations of mathematical information (Oxfam Education n.d.).

Exporting for the Future was a series of programs (Chapman 2003) designed for students in grades 9–12 to investigate Australia's place in international trade. While the program has been discontinued, the activities can be updated and used with students. Consider the example in Box 2 which illustrates how a critical mathematics perspective can be used to illuminate important aspects of everyday Australian life and the associated ethical/values-related sustainability and economic issues that impact this context.

Box 2

China is currently Australia's top trading partner. Can students:

i. read and interpret tables that represent changes in this trade relationship over time (e.g., $Am exports and imports, and percentage change over time)?
ii. create charts and graphs that visually represent the trends?
iii. explain reasons for the trends?
iv. identify and discuss social and environmental implications, and the links between them, of this trade relationship (e.g., using cheap Chinese labour, levels of pollution, exporting of Australia's mineral resources)

The *Fair Trade Association of Australia and New Zealand* has also prepared a range of interdisciplinary (mathematics, English, and social studies) teaching and learning resources. One example explores the banana trade (Fair Trade Association of Australia and New Zealand, n.d.) in which students use number and statistics to explore the impact of consumer purchase decisions on communities locally and around the world. They also gather, sort, and display data to explore patterns, variations, relationships, and trends relevant to the fair trade banana industry.

Working with practising teachers and the UK curriculum, *Amnesty International* developed a book for teaching and learning about human rights in the secondary mathematics classroom (Wright 2004). Examples include using the statistical concepts of mean, median, and mode to investigate differences in “average” wages by country, and using tables and charts to represent child labour data. Such activities allow for exploration of the impact of labour costs on the price of goods made in developing (third world) countries and exported to Australia. A recent high profile local case highlights how the ethical issues can be discussed in the classroom. It was brought to Australians’ attention that Sherrin, the Australian Football League’s football manufacturer, had illegally used child labour in the manufacturing process (see Levy et al. 2012). Mainstream Australians were outraged. Their ethical stance on the company’s practices impacted its corporate image, driving a clear message that such practices were unacceptable.

Examples from other Mathematics Content Domains

A cursory glance at some contemporary Victorian (Australia) mathematics textbooks suggests that efforts have indeed been made to provide students with examples clearly set in real world settings that are relevant or familiar to them.

Within the mathematical content area of ratio and rates, one example we found for grade 8 students related to the performance of a hybrid (petrol-electric) car (Vincent et al. 2007). Students were provided with data on the fuel consumption rate (litres per 100 km) of a Honda hybrid and asked to calculate a record breaking rate when the car was driven from Brisbane to Melbourne. A creative teacher could have followed up this activity by asking students to compare the rate found with the fuel consumption rate for a similar sized petrol-only car. Rhetoric questions such as, “Why doesn’t everyone drive a hybrid?” might provoke a discussion on relevant ethical, economic, and pollution issues.

On a very positive note, problems on projectile motion in the textbooks examined did not appear to involve military or other weapons. Occasionally, however, the contexts were somewhat contrived with questionable ethical overtones. In Box 3, one such example is presented.

> Box 3
> Students are told that a flowerpot is thrown from the third-storey of an apartment block and that the path followed is parabolic (the equation is given). A series of fairly traditional mathematical tasks follows – table of values, graph, horizontal distance travelled, domain and range of the function.
> [Modified from: Vincent et al., 2007]

Of concern with this example is that the issue of the flowerpot being *thrown* from the building appears to have been considered unproblematic. Yet we all know of cases where damage and/or death have resulted as a consequence of objects being thrown—for example, throwing stones/rocks from highway overpasses; Dowling (2013) reports one such incident.

Another example from Vincent et al. (2007) has great potential for teachers to initiate discussion on evidence to support or refute the effects of "climate change", an emotive and value-laden issue. In the introduction to the problem, the students are told that "[S]cientists believe that the decrease in population [of polar bears] is due to the earlier breaking up of the sea ice each summer over recent decades" (Vincent et al. 2007, p. 475). Students are asked to predict the population of polar bears in a given year based on linear mathematical modelling of the data presented and then to compare this with a predication based on an alternative exponential pattern (formula provided). The guidelines for teachers do not, however, suggest that students discuss this threat to the polar bear population or what other impacts earlier melting of sea ice might have on the planet.

Wright (2004) provides an opportunity for a statistical exploration of gender issues in society. Students are charged with examining a book or other resource, tabulating whether each image in the resource portrays a male or female, the occupation of the person, and what activity the person is engaged in. The data then have to be represented graphically and proportions calculated. Finally the students are asked "To what extent does the resource you have chosen show stereotypes?" (Wright 2004, p. 56), a prompt with the potential to challenge students' thinking about the issue. The reflective teacher has the opportunity to guide classroom discussion on gender equity and probe, raise awareness, and develop students' values on the issue. Since the students are in mathematics classrooms and the fields of mathematics and science remain perceived as male domains (Forgasz 2012), there is, at the same time, the opportunity to extend the discussion to incorporate and challenge these perceptions. Other statistical activities included in Wright's book encompass explorations to examine various other equity issues such as global inequalities, development indicators, life expectancy, refugee and asylum seeker status, literacy rates, poverty, and population changes. The final question to provoke thinking based on the statistical analyses of life expectancy data was "What factors do you think might cause a country to have high life expectancy or low life expectancy?" (p. 51).

Newspaper articles and advertisements are other rich sources of mathematical activities (Forgasz 1996). Carefully selected, the teacher can develop mathematical activities consistent with the curriculum, and promote discussion aimed at exposing underlying ethical and moral issues inherent to the context. In Box 4, an example from some years ago requires critical analysis of the source of the data—Triple J is a radio station with content and music aimed at young people—and who are responding to a phone-in listener poll on the issue of decriminalising marijuana.

Box 4

In a newspaper report in the Hobart Mercury in 1992 it was claimed that in response to a question on the decriminalisation of marijuana, 96% of the callers to the radio station Triple J were in favour (9924 out of 10,000 plus respondents), with only 389 believing that possession should remain a criminal offence.

In an article in the (Hobart) *Mercury on Saturday* research on ideal human body proportions for women and men was reported (Miranda 2002). A mathematics activity based on this article was developed (see Numeracy in the news, n.d.). The guidelines for teachers identified the damaging impact this article might have on young women, as well as the mathematical concept that could be used as the avenue to explore this issue:

> This article suggests that attractive bodies have legs 5% longer than average. This has major implications for body image. For most of us "attractive" proportions are impossible. Is it healthy, particularly for women to try to make up for leg shortfall with high heels? Because this article could actually create some angst in younger teenage girls we recommend using this for only grade 9/10's and using it as a springboard to look at data correlation. You will need a real Barbie Doll.

Based on the data in the article, students are required to find "height to average leg length" and "height to 'attractive' leg length" ratios. They then use class members' data to plot graphs of height versus leg length (for females and males) and find the gradients. The students are asked to use the graph to estimate what Barbie's height would be based on the leg length of the doll. Finally they are asked to "reflect on the impact of such articles… and dolls like Barbie in creating poor self body image concepts" (Numeracy in the news, n.d.). Mukhopadhyay and Greer (2001) described a similar mathematical activity used with pre-service and practising teachers who investigated what Barbie would look like if she were life size. The ensuing discussions were enlightening and effective, indicating that through mathematics, coupled with appropriate resources, teachers can promote critical thinking and raise students' awareness of issues that have potentially damaging social consequences.

Final Words

Teachers are responsible for deciding how to implement and enact an inclusive mathematics curriculum, and what resources and activities they will use with their students. While the resources and activities presented above certainly have the potential to probe students' existing attitudes, values, and understandings regarding important personal, local, and global issues, success in bringing these ideas into the classroom depends on teachers' willingness, knowledge, skills, motivations, confidence, and personal values and beliefs. Mathematics teachers will need to source appropriate contexts and activities and must remain vigilant to poor examples that have the potential to convey messages that are anathema to the development of appropriate ethical and moral understandings. If problem contexts such as the militaristic ones highlighted in Helen's story, the use of guns aimed at targets, or games of chance/gambling were brought into the classroom to exemplify relevant related mathematics concepts without the teacher challenging students' thinking regarding the implications of these contexts, there is the potential for inappropriate messages to be conveyed.

The Australian curriculum requires all teachers, including mathematics teachers, to develop their students' *ethical understanding*. Mathematics teachers should no longer be able to apply *ethical filtration,* or leave this task to teachers of other disciplines. But, how willing will mathematics teachers be to embrace this new responsibility? Will excuses be put forward to avoid embarking on this new challenge? To what extent do mathematics teachers accept interdisciplinary approaches to the teaching of mathematics? And, how prepared are they to open discussions on ethical issues to promote the development of their students as ethical, socially responsible, and informed citizens?

References

ACARA. (2011). The Australian curriculum: Mathematics v3.0. http://www.australiancurriculum.edu.au/Mathematics/Rationale. Accessed 24 Jan 2013.

ACARA. (May 2012a). The shape of the Australian curriculum. Version 3. http://www.acara.edu.au/verve/_resources/The_Shape_of_the_Australian_Curriculum_V3.pdf. Accessed 24 Jan 2013.

ACARA. (January 2012b). The Australian curriculum: General capabilities. Retrieved from http://www.australiancurriculum.edu.au/GeneralCapabilities/Pdf/Overview. Accessed 24 Jan 2013.

Australian Association of Mathematics Teachers (AAMT). (1998). Policy on numeracy education in schools. http://www.aamt.edu.au/Publications-and-statements/Position-statements/Numeracy-Education. Accessed 24 June 2014.

Bombardieri, M. (17 Jan 2005). Summers' remarks on women draw fire. *The Boston globe*. http://www.boston.com/news/local/articles/2005/01/17/summers_remarks_on_women_draw_fire/. Accessed 22 June 2014.

Chapman, S. (2003). *Exporting for the future. Trends in Australia's trade: Figuring out the figures*. Canberra: Austrade.

Commonwealth of Australia. (2009). Shape of the Australian curriculum: Mathematics. http://www.acara.edu.au/verve/_resources/Australian_Curriculum_-_Maths.pdf. Accessed 22 June 2014.

Danes, S. M. (1994). Parental perceptions of children's financial socialisation. *Financial Counselling and Planning, 5,* 127–146.

de Freitas, E. (2008). Critical mathematics education: Recognizing the ethical dimension of problem solving. *International Electronic Journal of Mathematics Education, 3*(2) (http://www.iejme.com/).

de Lange, J. (2003). Mathematics for literacy. In B. L. Madison (Ed.), *Quantitative literacy: Why numeracy matters for schools and colleges* (pp. 75–89). The National Council on Education and the Disciplines. http://www.maa.org/ql/whynumeracymatters.pdf.

de Meza, D., Irlenbusch, B., & Reyniers, D. (2008). *Financial capability: A behavioural economics perspective*. FSA Consumer Research Report 69. http://www.fsa.gov.uk/pubs/consumer-research/crpr69.pdf.

Dewey, J. (1930). *Human nature and conduct: An introduction to social psychology*. New York: The Modern Library Publishers.

Dewey, J. (2014 [1916]). *Democracy and education*. New York: Dover Publications.

Dowling, J. (2013). Man injured after rock thrown from freeway overpass. http://www.news.com.au/national/victoria/man-injured-after-rock-thrown-from-freeway-overpass/story-fn-do4cq1-1226579313894. Accessed 22 June 2014.

Duska, R., & Whelan, M. (1977). *Moral development: A guide to Piaget and Kohlberg*. Dublin: Gill & McMillan.

Ernest, P. (n.d.). Values and the social responsibility of mathematics. http://people.exeter.ac.uk/PErnest/pome22/Ernest%20%20Values%20and%20the%20Social%20Responsibility%20of%20Maths.doc. Accessed 22 June 2014.

Fair Trade Association (FTA) (n.d.). *All about bananas*. http://www.fta.org.au/school-resources.html. Accessed 22 June 2014.

Forgasz, H. (1996). Using newspapers as a resource for challenging activities. In H. Forgasz, T. Jones, G. Leder, J. Lynch, K. Maguire, & C. Pearn (Eds.) *Mathematics: Making connections* (pp. 102–107). Brunswick: Mathematical Association of Victoria.

Forgasz, H. (2005). Gender and mathematics: Re-igniting the debate. *Mathematics Education Research Journal, 17*(1), 1–2.

Forgasz, H. (2012). Gender issues and mathematics learning: What's new "down under"? In. L. Jacobsen (Ed.), *Mathematics teacher education in the public interest: Equity and social justice* (pp. 99–117). Charlotte: Information Age Publishing.

Frankenstein, M. (2008). In J. F. Matos, P. Valero, & K. Yasukawa (Eds.), Proceedings of the fifth international Mathematics education and society conference (pp. 261–271). Lisbon, Portugal: Centro de Investigação em Educação, Universidade de Lisboa and the Department of Education, Learning and Philosophy, Aalborg University. http://pure.ltu.se/portal/files/2376304/Proceedings_MES5.pdf.

Gellert, U., Jablonka, E., & Keitel, C. (2001). In B. Atweh, H. Forgasz, & B. Nebres (Eds.), *Sociocultural research on mathematics education. An international perspective* (pp. 57–73). Mahwah: Lawrence Erlbaum.

Gilligan, C. (1982). *In a different voice: Psychological theory and women's development*. Cambridge: Harvard University Press.

Gray, R., Bebbington, J., & McPhail, K. (1994). Teaching ethics in accounting and the ethics of accounting teaching: Educating for immorality and a possible case for social and environmental accounting education. *Accounting Education: An International Journal, 3*(1), 51–75.

Homer, P., & Kahle, L. R. (1988). A structural equation test of the value-attitude-behavior hierarchy. *Journal of Personality and Social Psychology,* 54(4), 638–646.

Ísaksson, A. (1979). Kohlberg's theory of moral development and its relevance to education. *Scandinavian Journal of Educational Research, 23*(2), 47–63.

Kohlberg, L. (1981). *Essays on moral development, Vol. I: The philosophy of moral development*. San Francisco: Harper & Row.

Kohlberg, L. (1987). *Child psychology and childhood education: A cognitive developmental view*. New York & London: Longman.

Levy, M., Webb, C., & Doherty, B. (2012). Sherrin child labour disgusting, Demetriou says. *The Age (online)*. http://www.theage.com.au/afl/afl-news/sherrin-child-labour-disgusting-demetriou-says-20120926-26k7p.html. Accessed 22 June 2014.

Lewis, A., & Scott, A. (2003). A study of economic socialisation: Financial practices in the home and the preferred role of schools among parents with children under 16. *Citizenship, Social and Economics Education, 5*(3), 138–147.

MCEETYA. (Dec 2008). Melbourne Declaration on Educational Goals for Young Australians. http://www.curriculum.edu.au/verve/_resources/National_Declaration_on_the_Educational_Goals_for_Young_Australians.pdf.

Miranda, C. (19 Jan 2002). Boffins agree our Kylie is just perfect. *Mercury on saturday* (p. 9) from http://www.simerr.educ.utas.edu.au/numeracy/teaching_activities/Health.html. Accessed 20 June 2014.

Mukhopadhyay, S., & Greer, B. (2001). Modelling with purpose: Mathematics as a critical tool. In B. Atweh, H. Forgasz, & B. Nebres (Eds.), *Sociocultural research on mathematics education. An international perspective* (pp. 295–311). Mahwah: Lawrence Erlbaum.

Mukhopadhyay, S., & Greer, B. (2002). Mathematics for socio-political criticism: The issue of gun violence. In S. C. Agarkar & V. D. Lale (Eds.), *Science, technology and mathematics education for human development* (Proceedings of the CASTME-UNESC)-HBCSE International Conference, Goa, India, February, 2001) (Vol. 2, pp. 195–199). Mumbai: Homi Bhabha Centre for Science Education, Tata Institute of Fundamental Research.

Numeracy in the News. (n.d.). http://www.simerr.educ.utas.edu.au/numeracy/default.html. Accessed 20 June 2014.

Oxfam Education. (n.d.). Water maths: Global statistics. http://www.oxfam.org.uk/education/resources/water_for_all/water/stats/index.htm. Accessed 22 June 2014.

Oxfam. (n.d.). Water for all. http://www.oxfam.org.uk/education/resources/water_for_all/. Accessed 22 June 2014.

Schoenfeld, A. (2003). Math wars. http://www.math.cornell.edu/~henderson/courses/EdMath-F04/MathWars.pdf. Accessed 22 June 2014.

Shim, S., Xiao, J. J., Barber, B. L., & Lyons, A. C. (2009). Pathways to life success: A conceptual model of financial well-being for young adults. *Journal of Applied Developmental Psychology, 30,* 708–723.

Shim, S., Barber, B. L., Card, N. A., Xiao, J. J., & Serido, J. (2010). Financial socialisation of first-year college students: The roles of parents, work and education. *Journal of Youth and Adolescence, 39,* 1457–1470.

Skovsmose, O. (n.d.). Critical mathematics education for the future. http://www.educ.fc.ul.pt/docentes/jfmatos/areas_tematicas/politica/CME_for_the_Future.pdf. Accessed 22 June 2014.

Skovsmose, O. (2008). Mathematics education in a knowledge market. In E. de Freitas & K. Nolan (Eds.), *Opening the research text. Critical insights and in(ter)ventions into mathematical education* (pp. 159–174). New York: Springer.

Skovsmose, O., & Valero, P. (2001). Breaking political neutrality: The critical engagement of mathematics education with democracy. In B. Atweh, H. Forgasz, & B. Nebres (Eds.), *Sociocultural research on mathematics education. An international perspective* (pp. 37–55). Mahwah: Lawrence Erlbaum.

Steen, L. A. (Ed.). (2001). *Mathematics and democracy: The case for quantitative literacy*. Princeton: National Council on Education and the Disciplines.

Tanner, L. N. (1997). *Dewey's laboratory school: Lessons for today*. New York: Teachers College Press.

Thomas, J. (2001). Globalisation and the politics of mathematics education. In B. Atweh, H. Forgasz, & B. Nebres (Eds.), *Sociocultural research on mathematics education. An international perspective* (pp. 95–112). Mahwah: Lawrence Erlbaum.

Vincent, J., Price, B., Tynan, D., Caruso, N., Romeril, G., Stillman, G., & Brown, J. (2007). *Maths worlds. year 9 VELS edition. Teacher edition*. Melbourne: Macmillan.

Vygotsky, L. (1978). *Mind & society*. Cambridge: Harvard University Press.

Vygotsky, L. (1986). *Thought and language*. Cambridge: The MIT Press.

Webley, P., & Nyhus, E. K. (2006). Parents' influence on children's future orientation and saving. *Journal of Economic Psychology, 27,* 140–164.

Wright, P. (2004). *Human rights in the curriculum: Mathematics*. London: Amnesty International.

Chapter 10
Valuing Diversity in Mathematics Pedagogy Through the Volitional Nature and Alignment of Values

Wee Tiong Seah and Annica Andersson

Introduction

Students make sense of and construct mathematical ideas in different ways, drawing upon their own unique experiences in life and in mathematics learning. Teachers' valuing of students' *diversity* of ideas fosters students' efficacy in learning mathematics and deepens students' mathematical understanding (Castellon et al. 2011; Schifter 2005; Zevenbergen et al. 2004).

This diversity in (mathematics) classroom discourses has taken on a different dimension in the last few decades or so, as cross-border human movements take place in arguably unprecedented levels, due to such developments as globalisation and regionalisation of trade and business activities, armed conflicts as well as low fertility rates in many developed countries. The composition of student ethnicities and cultures in classrooms across most countries has become very varied, and as such the diversity of student ideas and learning styles is greater than ever before.

However, what does the valuing of *diversity* in one's teaching practice mean? More often than not in the academic literature, this would mean adopting the view that nonmainstream students are struggling to learn mathematics, and that the teacher's role is to rescue them from failure and disengagement (see, for example, Ferguson 2009; Weber et al. 2010). Indeed, until recently, many mathematics education systems cater to diversity by creating differentiation and exclusivity, such as through giving different students different mathematics curricula, or through introducing different students to mathematics tasks of different difficulties.

In this chapter, however, we rethink diversity from a more empowering perspective. We turn our attention to how the diversity of student cultures (youth, class, ethnic, gender, linguistic, etc.) and hence how the diversity of students' mathematical

W. T. Seah (✉)
Faculty of Education, Monash University, Clayton, Australia
e-mail: WeeTiong.Seah@monash.edu

A. Andersson
Faculty of Mathematics and Science Education, Stockholm University, Stockholm, Sweden
e-mail: annica.andersson@mnd.su.se

A. Bishop et al. (eds.), *Diversity in Mathematics Education*,
Mathematics Education Library, DOI 10.1007/978-3-319-05978-5_10,

ideas can enrich students' mathematics learning experiences. What does it look like to be valuing this diversity? For some teachers, this may mean that they will encourage students to propose different solution approaches to any given mathematical task. For those teachers in this group who value *efficiency*, however, how might these values be in conflict? How do teachers negotiate about such professional and pedagogical valuing conflicts? Indeed, how might teachers plan their lessons so that they can be more proactive in teaching and modelling the valuing of *diversity* in the classroom discourse?

Yet, we also believe that students are aware of their capacities to adopt, resist or reject discursive positions. Given that there exists considerable within-class and within-school diversity of student cognitive and affective variables (Sullivan, Chapter 14, this volume), and in a context in which valuing is theoretically developed as a volitional act (a position which we will discuss below), classroom interactions between teachers and their students—and amongst students—represent sites of contestation and conflicts. This is inevitable, though we also do not desire for the domination of one particular (person's) goals and interests (Gutiérrez 2007). Quite clearly, the valuing of *diversity* is not about embracing different perspectives and ideas all at the same time.

We propose in this chapter that, through understanding valuing as volitional, the approach of aligning what students and teachers value facilitates the valuing of *diversity* in their respective mathematics classrooms, in ways which promote inclusivity and which optimise student learning of mathematics. In this chapter, we will be drawing upon observed episodes in mathematics lessons in schools to illustrate how this approach has been adopted successfully by teachers. In other words, we suggest that the ability to align values in one's classroom has been part of experienced teachers' craft. To achieve these, we will begin the chapter with a discussion of the nature of valuing relating to mathematics education, including an argument for values and valuing to be regarded as volitional in nature, representing commitment to a course of actions.

Valuing in Mathematics Learning and Teaching

Decisions and actions relating to the learning and teaching of mathematics in schools reflect directly what students and teachers value, and indirectly what is valued by parents and societies. Research into the role of values and valuing in mathematics learning and teaching began with Alan Bishop's proposal of three pairs of complementary values for 'Western' mathematics in the seminal book, 'Mathematical enculturation: A cultural perspective on mathematics education' (Bishop 1988). Our research findings in related studies (e.g. Andersson and Österling 2013; Andersson and Seah 2013; Seah 2005, 2011; Seah and Peng 2012) and our understanding of the academic literature (e.g. Bishop 1988, 1996; Clarkson et al. 2010a; Dede 2011; Hannula 2012; Lim and Kor 2010) led us to define values and valuing in mathematics education in the following way:

> Values are the convictions which an individual has internalised as being the things of importance and worth. What an individual values defines for her/him a window through which s/he views the world around her/him. Valuing provides the individual with the will and determination to maintain any course of action chosen in the learning and teaching of mathematics. They regulate the ways in which a learner's/teacher's cognitive skills and emotional dispositions are aligned to learning/teaching in any given educational context.

The extent to which a value is embraced and prioritised is responsive to one's environment and is thus not fixed. In other words, opportunities for values teaching in (mathematics) education exist across all school years. Whereas values may be absorbed when one is young (Court 1991), value priorities continue to be examined and evaluated throughout one's life in school and beyond. This may be seen in the valuing process that was conceptualised by Raths et al. (1987). Made up of three stages, that is, choosing, prizing and acting, the first stage is related to choosing freely and amongst several alternatives, after having thought about the consequences of adopting any one of these alternatives. It is this choosing activity that is stimulated by phenomena that allows for one's value priorities to be assessed. In a rather paradoxical way, this adds to the extremely internalised and stable (see Krathwohl et al. 1964) nature of values. In this way, there are opportunities for values such as *diversity* to be taught and to be reinforced in the school (mathematics) classroom.

Values through Mathematics

This teaching of values through the school subject of mathematics (values through mathematics) is one of two ways in which values in mathematics education can be considered. The other way would be the facilitation of mathematics pedagogy through the harnessing of values (mathematics through values).

Values are espoused and transmitted through the education system. Teachers' roles have always been involved with the teaching of values, even though such teaching is often implicit with even the teachers themselves often not aware of the process (Clarkson et al. 2000). Recent educational policies in countries such as Singapore (see Heng 2012) and Sweden (see Skolverket 2011) are in fact encouraging teachers to be more cognizant of their values-teaching role. Indeed, the Swedish School Law 4§ (Utbildningsdepartementet 2010, p. 2) states that

> the school education system support students in acquiring and developing knowledge and values. It shall promote all students' learning as well as a lifelong desire to learn. The education will also teach and establish respect for human rights and fundamental democratic values upon which the Swedish society is based. (Translation by Andersson)

After all, Veugelers and Kat (2000) observed that

> teachers cannot withdraw from showing the values that are important to them. In the cultural policy of the government and the school, teachers are even supposed to stimulate the development of specific values. (p. 11)

These educational values might be taught through dedicated school subjects labelled as 'civic education' and the like. Most, however, are transmitted via the teaching of other subjects, including the languages, history and the sciences. Thus, the valuing of *peace* or *diplomacy* can be taught in history lessons, while the valuing of *sustainability* or *precision* can be espoused by teachers of science. Mathematics too has the potential to be a medium through which such values are taught. For example, Andersson (2011a) as well as Andersson and Valero (in press) reported how values such as *global fairness* and *social justice* can be addressed in mathematics education. Sawatzki (2012) researched the teaching of financial values through mathematics lessons. Gutstein's (2006) work in Chicago suburban schools with mainly underprivileged Latino and black students also revealed rich possibilities to discuss relevant values in mathematics education.

Mathematics Through Values

The focus of this chapter, however, is not on the teaching of values (such as *diversity*) through mathematics. Rather, the intention is to examine how mathematics learning can be optimised through the harnessing of values. In particular, how might the mathematics learning experiences be made optimal through the valuing of *diversity* (of students' ideas and reasonings, and of students' experiences and cultures)?

Consider the following scenario that is rather commonly encountered in an Australian primary school class:

> Kathryn has been working with her students on place value. In this lesson, she had given each student a piece of A4-size card. The students were given time to draw (and decorate) their own favourite numeral on their given piece of card, after which three students were randomly selected to come to the front of the class with their cards. These three cards were held to their chests so that they were visible to the rest of the class. The three students' task was to arrange themselves in a straight line such that the three-digit number formed is the largest possible in magnitude. This process was then repeated with a few other groups of three randomly-selected students.

The expectation for students in the class to listen to the explanations of their three peers up in front represents the teaching of *respect* or *democracy*. In most if not all cultures, either of these values would be a desirable one to introduce to students. Yet, in this example of 'values through mathematics', neither of these values is related to mathematics or to mathematics pedagogy alone, and either may be introduced to students in any other lesson at school.

The scenario above also illustrates the introduction of at least two other values, both of which however, are related to mathematics or mathematics education in a unique way. The students' explanations exemplify the valuing by the teacher of *openness* (see Bishop 1988); each group of three students needed to defend publicly their choice of a three-digit number which satisfied the task criteria. This value expresses the spirit of mathematicians, in whose professional lives new theorems and formulae are accepted only after these are disseminated and examined within the scientific circle.

Kathryn (the teacher) could have taught place value without facilitating the group activity outlined in the quote above. The fact that she did reflects her valuing and portrayal of *fun*. That is, Kathryn was making use of this valuing to make mathematics learning enjoyable, so as to optimise the learning experience for her students. Her portrayal of *openness* through this activity shared the same objective. Through this portrayal, students can appreciate the benefits of peer communication and discussion in facilitating understanding. For some of them, the valuing of *openness* may enable them to find the subject less threatening, less impersonal and, hopefully, more rewarding.

Thus, we have here a classroom scenario in which the representation and possibly the inculcation of the mathematical value *openness* and of the mathematics educational value *fun* served to promote more engaged and 'effective' mathematics learning amongst students. At the same time, the learning of mathematics is made more inclusive to all students. Values were portrayed not for their own sake, but with the objective to optimise mathematics learning and to make it open to all.

This scenario has also explicated what Bishop (1996) proposed as the three categories of values that function in the school mathematics classroom, these being mathematical (the valuing of *openness* in the example), mathematics educational (*fun*) and general educational (*respect*).

The category of mathematical values refers to the convictions that are emphasised in the discipline of mathematics. In relation to 'Western' mathematics, for example, Bishop (1988) had proposed three pairs of complementary mathematical values, namely *rationalism* and *objectism*, *control* and *progress*, as well as *mystery* and *openness*.

Mathematics educational values are expressed through the pedagogical practices of the subject in schools. They are understandably situated in the sociocultural context of different education systems, and as such, the range of these values can be extensive. Examples of this category of values include *ICT, practice*, *ability* and *effort*—one or more of these, amongst many others, may be valued by any mathematics teacher in his/her pedagogical practice.

Consider yet another scenario in the mathematics classroom:

> Sheridan teaches in a Grade 5 class in a primary school in Melbourne, Australia. There are many migrant children amongst her students. A few months ago she was introducing the algorithm for multiplying multiple-digit numbers such as 24×37. Towards the end of her demonstration and explanation, a migrant student raised her hand to offer 'a quicker method, Miss Sheridan'. This student then proceeded to show the class the lattice method of multiplication, which her dad had taught her at home. The student also shared that her dad told her that he had learnt this quicker way of multiplying in his primary school days in his home country.

The reader should be able to identify with incidents such as the one faced by Sheridan, involving other alternative algorithms or, indeed, other mathematical topics. In a critical incident (see Tripp 1993) such as this, Sheridan's response is crucial in shaping the students' understanding of the nature of mathematics and their feelings of what it means to be learners of mathematics. That is to say, Sheridan's response will not affect just the particular student who raised her hand. For Sheridan, this was the first time she was shown the lattice method, and she was as amazed as some of her students at seeing how the product was efficiently and accurately computed. Since the student did not know what the method was called, Sheridan was not able to check it out on the laptop she had with her in class on-the-spot. Yet, she wanted to capitalise on that moment in the lesson when the students were eagerly watching the lattice method unfolding before their eyes, and so Sheridan asked the class to work in their groups of threes to 'test this method out'. She made use of the five minutes she had to compare the lattice method with the 'textbook method' of multiplying multiple digits. Being able to see similarities between the two, it gave her the confidence to ask the class to report their findings. Sheridan made use of the opportunity to ask the students to check the products they obtained with the lattice method against the corresponding answers they would have obtained using the method she taught them at the beginning of the class. She ended the lesson promising her students that she would look for the name of the alternative algorithm.

In this scenario, Sheridan's response to the migrant student—and to her class more generally—reflected her valuing of *diversity* and *openness*. Sheridan was modelling the importance and worth she dedicated to the diversity of mathematical ideas amongst her students. She was ready to entertain alternative ways of multiplying, and for this reason she did not select the easy way out by telling the student to ignore the mathematics she already knew, and to focus instead on learning the 'right' way of multiplying numbers. *Diversity* was being presented here as a general educational value in the Australian society, corresponding as it does with one of the nine values for Australian schooling, namely, *understanding, tolerance and inclusion*. Valuing this means to be 'aware of others and their cultures, accept diversity within a democratic society, being included and including others' (Australia Department of Education Employment and Workplace Relations 2011, np). This general educational value is also promoted in many other cultures too. For example, through the Swedish School Law 8§ (Utbildningsdepartementet 2010), Swedish teachers are made aware that 'all children and young people, regardless of gender, place of residence, and socioeconomic status, have equal access to education in public schools' (translation by Andersson).

The mathematical value of *openness* was espoused by Sheridan as well. Instead of simply accepting or rejecting the student's offer of a more efficient method so that she could get on with her planned pedagogical activities, Sheridan facilitated a group-based investigation to verify that the lattice method works for different multiple-digit multiplication situations. In so doing, she demonstrated how knowledge can be democratised through verifications and (student) explanations, which in Bishop's (2008) view is the defining feature of the valuing of *openness*.

Before we move on, it is worth noting that *diversity* can also be regarded as a mathematics educational value. That is, *diversity* may also be valued as part of teachers' pedagogical decisions or activities. The Australian Curriculum, for example, encourages teachers to structure in their lesson planning for students who are learners of English as an additional language or dialect (EAL/D), 'additional time and support, along with teaching that explicitly addresses their individual language learning needs' (ACARA 2011, np).

Volitional Nature of Values

Mathematics educational research regarding values has previously been considered as part of the affective tradition. Bishop's (1999) definition that 'values in mathematics education are the deep affective qualities which education fosters through the school subject of mathematics' (p. 2) reflects this stance. The affective domain of the Taxonomy of Learning Objectives (Krathwohl et al. 1964) also positioned valuing as the emotional outcome that develops from attitudes and beliefs.

The valuing process (Raths et al. 1987), however, hints at the involvement of a certain degree of cognition as values are being internalised. The stage of choosing, for example, involves cognitively based considerations amongst alternatives to enable a choice to be made. The way that civic and moral values are taught in schools in some countries also demonstrates the cognitive nature of the acquisition of a value. These teaching approaches include the values clarification exercise method (see Simon and Clark 1975), a pedagogical approach which invites students to state their respective positions with regard to some given scenario, and to explain (and indeed clarify) their choices. The cognitive involvement is especially evident during this values clarification stage of the lesson, as students are made aware of what they value and of the extent to which this valuing is shared by peers. Also, the internalisation of a value by putting it into action is often defensible by an individual—another sign of valuing as involving cognition rather than it being an emotional response alone.

But is it sufficient to consider values and valuing as being cognitive/affective variables? How does this affect the ways in which we understand this construct, and how can we better utilise it to optimise mathematics learning and teaching for all students? In what ways might a new understanding of the nature of values and valuing contribute towards 'a process of identifying any barriers within and around the school that hinder learning, and reducing or removing these barriers' (UNESCO 2001, p. 16), and in so doing, fostering inclusive education in schools?

Consider, for example, a student who has been taught the range of mathematical problem-solving strategies, and who has developed confidence and efficacy in engaging with mathematical problem-solving. Having the cognitive skills and emotional dispositions, however, does not mean that the student will necessarily engage in—or value—problem-solving. The student might value *creativity* instead, and having found the problem-solving questions actually quite routine or predictable, she/he may decide to devote attention and engagement in other school subjects which allow him/her to display *creativity*.

As another example, the valuing of *achievement* in school mathematics does not necessarily mean that an individual is yet to possess the cognitive skills to achieve what is being aimed for, or that she/he is emotionally positive about it. Nevertheless, she/he attaches importance to *achievement*, such that this valuing drives him/her to do what is needed to attain it, thus actualising it. Valuing *achievement* may prompt the individual to seek extra assistance (such as home tuition), and this valuing may sustain his/her efforts to do so even if she/he might not be interested or motivated in the tasks involved. Indeed, we feel that this might underlie the relatively high performance of East Asian students in international comparative tests, despite the common schooling experience they have with peers from other cultures, and despite these students in Asia not liking mathematics generally (OECD 2013). It might be said that the students' cultural values were responsible for, or at least contributed to, their valuing of *achievement*. Thus, we argue that for the East Asia students, these cultural values—as well as the mathematical and mathematics educational values that had been internalised over the years as learners—provide them with the drive and the will to succeed and to do well in school mathematics.

The philosopher Ayn Rand wrote that 'a being of volitional consciousness has no automatic course of behaviour. He [*sic*] needs a code of values to guide his actions' (1961, p. 97). Thus, values can be seen as variables which are action based. This is not to say, however, that values are always expressed in the form of actions. Rather, the potential for action is the basis for valuing. Whether it is expressed in action or not depends on the context. This can be seen in Andersson (2011b), in which the upper secondary student participants who had indicated that they disliked—even hated—mathematics, clarified that their stories of mathematics learning experiences were connected to the context in which they were told. That is, the students' stories and actions for learning mathematics changed as the contexts evolved. As volitional variables, values have both cognitive and affective components as well. The cognitive components are visible through the choosing dimension of the valuing process (Raths et al. 1987). That valuing also has an affective dimension can be seen in the way we often find ourselves embracing what we value in a passionate way, supported by associated emotions, attitudes and beliefs.

Values: Motivation or Will?

From Hannula's (2012) theoretical standpoint within the motivational research area, values as a volitional construct can be regarded as a motivational agent. Motivation initiates and directs action, but it may not be responsible for sustaining the action. Yet, whilst values guide an individual to choose a course of action, they appear to do more beyond this function. A teacher who values *exploration*, for example, will not only be motivated to plan and deliver his/her lessons in ways which include student investigation tasks and group discussion opportunities. Equally important, this valuing places so much emphasis on the worth of exploration that the teacher's actions will serve to sustain this pedagogical approach should obstacles arise. Thus, a parent may question the wisdom of 'wasting' instructional time in allowing students to 'discuss mathematics'. If values are mere motivational forces, such a parent intervention may affect the teacher's motivation to continue to facilitate exploration amongst the students. Yet, as a value, the teacher will respond in ways which help him/her to continue to value *exploration* in his/her professional practice. Such a response might include a reply to the parent arguing for the case of inculcating student valuing of *exploration*, or explicit explanations to students of the advantages to their mathematics learning of their opportunities to explore strategies. In other words, an individual not only acts on what she/he values, but defends what she/he values as well.

As explained by Kivinen (2003),

> there is a distinguishing line between volition and motivation. Volition promotes the intent to learn and protects the commitment and concentration from competing action tendencies and other distractions …. For example, a student may be motivated to read a book in the evening. He or she is more or less motivated to do so. The student takes the book and starts to read (motivation has done its work). Volitional processes (will) keep him or her reading, in spite of the fact that there is an interesting football match on TV. (pp. 26–27)

In other words, as a volitional variable, values (in mathematics education) not only motivate and guide decisions and actions, but they also provide the individual with the will and determination to maintain this course of action in the face of competing actions and obstacles.

It is this second aspect of the volitional characteristic of values which gives them their characteristic soul. The importance attached to a value is reflected not just through its regulating action, but also through the ways in which it provides the individual with a will and determination to continue valuing it. A country's valuing of *freedom* and/or *justice* may lead to decisions about fighting a war half a globe away, but it is how this valuing is sustained in the face of public protests and academic doubts that allows for the effects of this valuing to be felt. In the context of the school mathematics classroom, then, there is every hope that teacher harnessing of the valuing of *diversity* of students' mathematical ideas and reasoning, and of *diversity* of students' ethnicities and ways of thinking, will stimulate a sense of greater inclusivity in the classroom.

The Significance of Valuing When Facilitating Mathematics Learning and Teaching

Thus, valuing provides one with the will and determination to act in particular ways. This may involve the modification of related beliefs and other emotional constructs, as well as the choice of mental strategies and decisions deployed to 'do mathematics'. In other words, what is valued regulates cognitive processes and affective modes. We now provide some examples to highlight the significance of values in mathematics learning and teaching.

It is often observed in some 'Western' societies (see, for examples, Byun and Park 2012; Wei and Eisenhart 2011) that Asian students, especially East Asian students, perform better than their peers in school mathematics. These students would, however, be attending the same schools as their peers from other ethnic backgrounds. That is, they and their peers would have been taught by the same teachers, have performed similar activities during mathematics lessons, be expected to do the same homework and have sat for the same assessment tasks. They would also have experienced the same classroom learning environment and conditions, as well as the same external motivational factors. Given these same educational opportunities, then, why do East Asian students perform better in school mathematics? For these East Asian students, what are the underlying cultural values conveyed by parents and the wider society which might support the students' mastery of cognitive skills and development of affective dispositions in the school mathematics classroom?

At the same time, again considering the case with East Asian students, there is no conclusive evidence of any correlation between affect and performance. For example, even though Grootenboer and Hemmings (2007) found that affective factors such as beliefs and confidence were associated significantly with mathematical performance, they were not predictive of performance. Indeed, students from Singapore and Hong Kong (top performing countries in international comparative assessments) can often be heard expressing their dislike of or stress with school mathematics. Quite clearly, the level of mathematical wellbeing (Clarkson et al. 2010b) is not high for these high-performing students.

The results of PISA 2003 told a similar story. In PISA 2003, students' interest in and enjoyment of mathematics were highest in Tunisia, followed by Indonesia, Thailand, Mexico and Brazil. Yet, students from these five countries also occupied the last five spots in the country ranking by performance (OECD 2004). These observations suggest that favourable affective dispositions in students may not be sufficient to bring about 'effective' learning and performance in school mathematics. Some factors beyond the cognitive and affective ones are at play. Several reports (e.g. Leung 2006; Wei and Eisenhart 2011) have made reference to culturally based values in mathematics education. Askew et al. (1997) appeared to stop short of naming these as the factor regulating 'effective' teaching, referring instead to teachers 'believing in the importance of' (p. 4) particular pedagogical practices in their mathematics teaching repertoire. It is these variables of importance and worth which constitute the focus of this chapter.

Values Alignment in the Mathematics Classroom

We are interested in this chapter to make use of a theoretical stance, understanding values and valuing as volitional with the purpose to support teacher modelling and teaching of *diversity* in the classroom discourse. We are mindful that teachers and their respective students come to class with their own aspects of valuing. The decisions and actions of teachers and students in the mathematics classroom reflect their respective valuing. How do teachers (and students) go about negotiating the differences that inevitably exist, so as to facilitate learning of the subject? Indeed, in Australia for example, teachers' unmet professional development needs are centred round ways of planning for diverse student needs and capabilities (see Panizzon,

Chapter 3). Why is it that teachers whose practice may be fruitful in one classroom may not find him/herself equally 'effective' in another?

Given the volitional nature of values, it is reasonable to argue that any expectation by teachers (students) for others in class to share their valuing automatically is heading for failure. In order to maintain a functioning classroom environment amidst the range of values present, teachers and students will want to see one another's values to be aligned and in harmony. In fact, if we perceive the classroom as an organisation that is dedicated to learning mathematics, then Senge's (2006) five disciplines of learning organisations would foreground values alignment as a crucial attribute. In particular, one of them, personal mastery, would require the teacher and student to align her/his values with those of the class. Also, the discipline of building shared vision also calls for aligned values in an organisation in order for the shared vision of the future to be co-created. Thus, central to our discussion in this chapter is the notion of values alignment:

> Building … values alignment is about providing a cooperative and collaborative process whereby the members of the organisation can develop strategies, systems and capabilities that not only support those values that have previously been clarified as being essential for the ultimate success of the group as a whole but also are supported by the majority of the people within the group as acceptable guidelines for directing their behaviour (Henderson and Thompson 2003). (Branson 2008, p. 383)

Although values alignment may be a concept borrowed from the field of business administration (see, for example, Branson 2008), its appropriateness in accounting for the professional practice of teachers negotiating about value differences is supported by the observation that in both a business organisation and school organisation, there is often a desire amongst the employers/management and the teachers/school for a shared goal. In addition, the employees and the students respectively would subscribe to these goals to different degrees. The value of values alignment lies in the observation that

> all relationships—between one person and another, between the present and the future, between customer and product, a team and its goals, a leader and a vision—are claimed to be strengthened by aligned values. (Branson 2008, p. 381)

Thus, for a teacher, being able to facilitate values alignment between what she/he values and what his/her students value promises to strengthen the relationships, and is one of the keys to nourishing teaching and learning practices. Indeed, we propose that teachers are seen as being 'effective' in different classrooms because they are resourceful and creative enough to attain values alignment in whatever classroom situation they find themselves.

Even though the teacher may take a leading role in facilitating values alignment within his/her classroom, student agency might not be lost (Andersson and Valero, in press). A key assumption in our work is that all students are agentic in either adopting or resisting/rejecting discursive positions. Students are explicitly and implicitly providing their teachers with feedback about their learning, both consciously and subconsciously, just as teachers provide similar feedback to their respective students. These feedback data should allow the teacher to monitor the extent to which the values being negotiated are accepted by their students, and/or the extent to which they are being shaped by the students in turn. We can see examples of these in the two cases provided below.

It needs to be reminded that values alignment is not about facilitating a classroom situation in which everyone or most people subscribe to the same values. Values alignment is different from values inculcation; in fact, it is doubtful that values inculcation—or even an expectation of it—would ever be accepted by students, parents and teachers in most classroom situations today. Rather, values alignment facilitates the coexistence of different values as these are held by different people interacting in any given context. In so doing, students perceive that their knowledge, skills and dispositions are valued. They feel valued and inclusive in relation to their learning of mathematics. That is to say, values alignment acts as an agent for promoting inclusivity in the mathematics classroom.

Case 1

> Michael (a pseudonym) was a mathematics teacher in a secondary school. He had noticed that his Grade 10 students had been unwilling to work with concrete manipulatives such as geoboards and pattern blocks. 'These are for young kiddies, sir!' they would say. Yet, Michael felt that learning is more effective when students are able to visualise the relevant concepts. Michael's students are now exploring and understanding geometrical concepts through software programs, such as dynamic geometry software.

In this case, Michael's use of concrete manipulatives reflects his valuing of *visualisation*. However, this teaching approach was resisted by his students whose values were not aligned with the image of teenagers 'playing with blocks'. There was a potential here of a value conflict between Michael and his students, which could possibly result in the students being disengaged with his lessons. What Michael did to resolve the potential value conflict was his redefining of what he and his students value, coming to an understanding that in effect, his valuing of *visualisation* was underlied by a valuing of *exploration*. This was crucial, since the students' values were aligned with *exploration* as well; it was just that they did not want to feel like small kids playing with blocks and teddy bears. By redefining his valuing of visualisation with the use of digital learning technologies, Michael was able to plan and execute his lessons such that the dynamic geometry software provided the students with opportunities to explore—and thus visualise—the relevant geometrical ideas and concepts in a form that is now aligned with what the students value. Michael's valuing of *visualisation* had given him the will to resolve the value difference situation in ways which still allow for student visualising to take place, only that the means of actualising this valuing were now accepted by the teenage students, who were understandably trying to look more adult-like and doing adult tasks. For his students, their positive response to the ICT use was an endorsement of their common valuing of *exploration* as well.

In this instance, then, values alignment was achieved through Michael's volitionally redefining what he values such that its expression now is aligned with what his students value.

Case 2

> Diane was an immigrant secondary school mathematics teacher from Canada. When one of her students answered one of her questions by saying 'just chuck in c, just chuck in the c', she responded that he was being too casual with his use of mathematical language. Diane's

> own mathematics learning experience in Canada had instilled in her a valuing of the formality in mathematics, a tradition which she felt needed to be upheld. Thus she would have preferred her students to talk about 'adding the constant, c'.
> Yet, Diane was deeply aware and concerned that she was teaching a 'weak' class, and that meant that it would not be wise to get 'too caught up in those formal, scary things'. She was mindful that for these students, a valuing of fun would be a key motivator for them. As such, she made a conscious effort to 'sacrifice "plus c" for "chuck in a c"'.

Here, Diane had realised that 'pushing' her students to share her valuing of *formality* and to use formal terminology would be counter-productive. This group of students needed first and foremost to be able to be interested enough in the subject, and to develop some confidence to acquire the skills and concepts required of them. The students' valuing of *fun* was a volitional force which supported the cognitive and affective growth that they needed. Diane's understanding of this, and her subsequent re-prioritisation of her valuing of *formality* and *fun*, resulted in values alignment between herself and her students. This re-prioritisation of Diane's values was evident when she talked about the relative importance of notations/formality and understanding/enjoyment, and how it would be her willing sacrifice to interchange the order of priority for the sake of facilitating her students' learning.

In this second case, values alignment was achieved when one group of people (in this case, Diane the mathematics teacher) in interaction re-prioritises what they value, such that there is now a common valuing in the whole group. If valuing as a volitional variable means that the individual subscribing to particular values will sustain the valuing at all costs, this feature is not violated here: Diane still values *formality*. However, she also shares students' valuing of *fun*, and a re-prioritisation between these two values within herself has resulted in an alignment of what she and her students value.

Concluding Ideas

The writing of this chapter was motivated by our interest in examining how we might support teachers' valuing of *diversity* which we believe in turn optimises students' learning experience. We have theoretically argued that values and valuing might be volitional in character. The lesson scenario involving Sheridan provides us with an example of how this valuing might look like in practice.

Yet, the mathematics classroom is a place where the different values of teachers, students and indirectly others come together in intersection, resulting in value differences and value conflicts. If teachers' espousal of *diversity* is key to facilitating an inclusive learning environment and to optimising the learning experiences of all students, it is important for teachers to be able to negotiate these value differences and value conflict situations in their respective classes.

We propose in this chapter that the awareness and alignment of present values in the different educational and classroom contexts can empower teachers to achieve this. In discussing this approach, we highlighted the will aspect of values and valuing. This drive to maintain the course of action is evident in both Michael and Diane, whose respective actions at aligning their own values with their students' were aimed at optimising student learning. Michael's redefining his valuing as *exploration*, and Diane's prioritising of the valuing of *fun* over that of *formality*, are two examples of teacher strategies that might be adopted to bring about values alignment.

Diane's case shows how *fun,* as it was valued by her and her students, was also actualised in the norms of the class. In the process, Diane's valuing of *formality* was given a relatively lower priority. For Diane, the different values (*student learning*, *fun* and *formality*) appeared to be internalised to different degrees within herself. While each of these values is volitional, it is not possible for all three values to be emphasised to the same degree. Diane's overriding emphasis on (student) *learning* thus guided her to prioritise *fun* over *formality* in order to embody this overriding emphasis.

Values alignment in Michael's case, however, did not involve the teacher and students embracing the same values. Michael's decision to make use of dynamic geometry software in his lessons allowed him to express his valuing of *visualisation* in another form, one that does not involve the use of concrete manipulatives. In so doing, this expression of the teacher valuing of *visualisation* supports and is in alignment with what the students valued, underlied by the common valuing of *exploration* amongst Michael and his students.

The challenges for teachers, however, are firstly, the extent to which they are aware of what they personally value in relation to the mathematics subject, to mathematics pedagogy and to general educational aims. Secondly, there is also the challenge of being more 'effective' at facilitating values alignment within the teachers' own classrooms. These challenges are by no means unrelated: If teachers are not able to articulate their personal convictions, their values alignment experience to facilitate student learning will remain tacit. Thus, teacher capacity to actualise values alignment between herself/himself and her/his students go a long way towards acknowledging students' cultures, knowledge, skills and dispositions, thereby contributing to *diversity* in mathematics learning and teaching in ways which are inclusive and empowering.

References

Andersson, A. (2011a). A 'curling teacher' in mathematics education: Teacher identities and pedagogy development. *Mathematics Education Research Journal, 23*(4), 437–454. doi:10.1007/s13394-011-0025-0.

Andersson, A. (2011b). *Engagement in education: Identity narratives and agency in the contexts of mathematics education*. Doctoral thesis, Aalborg University, Aalborg, Denmark, Uniprint.

Andersson, A., & Österling, L. (2013). Measuring immeasurable values. In A. M. Lindmeier & A. Heinze (Eds.), *Proceedings of the 37th Conference of the International Group for the Psychology of Mathematics Education* (Vol. 2, pp. 17–24). Kiel: PME.

Andersson, A., & Seah, W. T. (2013). Facilitating mathematics learning in different contexts: The values perspective. In M. Berger, K. Brodie, V. Frith, & K. L. Roux (Eds.), *Proceedings of the 7th international mathematics education and society conference* (Vol. 1, pp. 193–202). Capetown: MES 7.

Andersson, A., & Valero, P. (in press). Negotiating critical pedagogical discourses. Stories of contexts, mathematics and agency. In P. Ernest & B. Sriraman (Eds.), *Critical mathematics education: Theory and praxis*. USA: Information Age Publishing.

Askew, M., Brown, M., Rhodes, V., Johnson, D., & William, D. (1997). *Effective teachers of numeracy. Final report*. London: King's College.

Australian Curriculum Assessment and Reporting Authority [ACARA]. (2011). Australian curriculum: Students for whom English is an additional language or dialect. http://www.australiancurriculum.edu.au/StudentDiversity/Students-for-whom-English-is-an-additional-language-or-dialect. Retrieved 16 June 2013.

Australia Department of Education Employment and Workplace Relations. (2011). National framework: Nine values for Australian schooling. http://www.valueseducation.edu.au/values/val_national_framework_nine_values,14515.html. Retrieved 11 June 2013.

Bishop, A. J. (1988). *Mathematical enculturation: A cultural perspective on mathematics education*. Dordrecht: Kluwer.

Bishop, A. J. (1996 June 3–7). *How should mathematics teaching in modern societies relate to cultural values—some preliminary questions*. Paper presented at the seventh Southeast Asian conference on mathematics education, Hanoi, Vietnam.

Bishop, A. J. (1999). Mathematics teaching and values education: An intersection in need of research. *Zentralblatt fuer Didaktik der Mathematik, 31*(1), 1–4.

Bishop, A. (2008). Teachers' mathematical values for developing mathematical thinking in classrooms: Theory, research and policy. *The Mathematics Educator, 11*(1/2), 79–88.

Branson, C. M. (2008). Achieving organisational change through values alignment. *Journal of Educational Administration, 46*(3), 376–395.

Byun, S-Y., & Park, H. (2012). The academic success of East Asian American Youth. *Sociology of Education, 85*(1), 40–60.

Castellon, L. B., Burr, L. G., & Kitchen, R. S. (2011). English language learners' conceptual understanding of fractions: An interactive interview approach as a means to learn with understanding. In K. Téllez, J. N. Moschkovich, & M. Civil (Eds.), *Latinos/As and mathematics education: Research on learning and teaching in classrooms and communities* (pp. 259–282). Charlotte: Information Age Publishing.

Clarkson, P., Bishop, A. J., FitzSimons, G., & Seah, W. T. (2000 July 31–August 6). *Life-long learning and values: An undervalued legacy of mathematics education?* Paper presented at the 9th international congress on mathematical education, Tokyo/Makuhari, Japan.

Clarkson, P., Bishop, A., & Seah, W. T. (2010a). Mathematics education and student values: The cultivation of mathematical well-being. In T. Lovat & R. Toomey (Eds.), *International handbook on values education and student well-being* (pp. 111–136). NY: Springer.

Clarkson, P., Seah, W. T., & Bishop, A. (2010b). Mathematics wellbeing and teacher education. In R. Toomey, T. Lovat, N. Clement, & K. Dally (Eds.), *Teacher education and values pedagogy: A student wellbeing approach* (pp. 179–194). New South Wales: David Barlow Publishing.

Court, D. (1991). Studying teachers' values. *The Clearing House: A Journal of Educational Strategies, Issues and Ideas, 64*(6), 389–392.

Dede, Y. (2011). Mathematics education values questionnaire for Turkish preservice mathematics teachers: Design, validation, and results. *International Journal of Science and Mathematics Education, 9,* 603–626.

Ferguson, S. (2009). Same task, different paths: Catering for student diversity in the mathematics classroom. *Australian Primary Mathematics Classroom, 14*(2), 32–36.

Grootenboer, P., & Hemmings, B. (2007). Mathematics performance and the role played by affective and background factors. *Mathematics Education Research Journal, 19*(3), 3–20.

Gutiérrez, R. (2007). (Re)defining equity: The importance of a critical perspective. In N. S. Nasir & P. Cobb (Eds.), *Improving access to mathematics: Diversity and equity in the classroom* (pp. 37–50). NY: Teachers College Press.

Gutstein, E. (2006). *Reading and writing the world with mathematics: Toward a pedagogy for social justice*. NY: Routledge.
Hannula, M. S. (2012). Looking at the Third Wave from the West: Framing values within a broader scope of affective traits. *ZDM, 44,* 83–90.
Henderson, M., & Thompson, D. (2003). *Values at work: The invisible threads between people, performance and profit*. Auckland, New Zealand: Harper Collins.
Heng, S. K. (2012). Opening address at the 2012 Ministry of Education Work Plan Seminar. http://www.moe.gov.sg/media/speeches/2011/09/22/work-plan-seminar-2011.php. Retrieved 11 Jan 2013.
Kivinen, K. (2003). *Assessing motivation and the use of learning strategies by secondary school students in three international schools*. Unpublished PhD dissertation, University of Tampere, Finland.
Krathwohl, D. R., Bloom, B. S., & Masia, B. B. (1964). *Taxonomy of educational objectives: The classification of educational goals (Handbook II: Affective domain)*. New York: David McKay.
Leung, F. K. S. (2006). Mathematics education in East Asia and the West: Does culture matter? In F. K. S. Leung, K.-D. Graf, & F. J. Lopez-Real (Eds.), *Mathematics education in different cultural traditions: A comparative study of East Asia and the West* (pp. 21–46). Springer.
Lim, C. S., & Kor, L. K. (2010). Mathematics classroom practice of 'excellent' teachers: What can we learn? In Yoshinori Shimizu, Y. Sekiguchi, & K. Hino (Eds.), *Proceedings of the 5th East Asia Regional conference on mathematics education* (Vol. 1, pp. 118–126). Tokyo: Japan Society of Mathematics Education.
Organisation for Economic Co-operation and Development [OECD]. (2004). *Learning for tomorrow's world: First results from PISA 2003*. Paris: Organisation for Economic Co-operation and Development.
Organisation for Economic Co-operation and Development [OECD]. (2013). *PISA 2012 Results: What students know and can do—student performance in mathematics, reading and science* (Vol. 1). Paris: PISA, OECD.
Rand, A. (1961). *For the new intellectual: The philosophy of Ayn Rand*. NY: Signet.
Raths, L. E., Harmin, M., & Simon, S. B. (1987). Selections from 'values and teaching'. In J. P. F. Carbone (Ed.), *Value theory and education* (pp. 198–214). Malabar: Robert E. Krieger.
Sawatzki, C. (2012). *Investigating the role of financial values and attitudes in financial literacy teaching and learning*. Paper presented at the 35th annual conference of the mathematics education Research Group of Australasia.
Schifter, D. (2005). Engaging students' mathematical ideas: Implications for professional development design. *Journal for Research in Mathematics Education Monograph* (Vol. 13). Reston: National Council of Teachers of Mathematics.
Seah, W. T. (2011). Effective mathematics learning in two Australian Primary classes: Exploring the underlying values. In B. Ubuz (Ed.), *Proceedings of the 35th conference of the international group for the psychology of mathematics education* (Vol. 4, pp. 129–136). Ankara: PME.
Seah, W. T., & Peng, A. (2012). What students outside Asia value in effective mathematics lessons: A scoping study. *ZDM: The International Journal on Mathematics Education, 44,* 71–82.
Senge, P. M. (2006). *The fifth discipline: The art and practice of the learning organization* (2nd ed.). London: Random House.
Simon, S. B., & Clark, J. (1975). *More values clarification*. San Diego: Pennant Press.
Skolverket. (2011). *Läroplan, examensmål och gymnasiegemensamma ämnen för gymnasieskola 2011*. Stockholm: Fritzes förlag.
Tripp, D. (1993). *Critical incidents in teaching: Developing professional judgement*. London: Routledge.
UNESCO. (2001). *Understanding and responding to children's needs in inclusive classrooms: A guide for teachers*. Paris: UNESCO.
Utbildningsdepartementet. (2010). *Svensk författningssamling 2010:800*. Stockholm: Utbildningsdepartementet.

Veugelers, W., & Kat, E. (2000 April 24–28). *The teacher as a moral educator in secondary education: The student perspective*. Paper presented at the 81st annual meeting of the American Educational Research Association, New Orleans, LA.

Weber, K., Radu, I., Mueller, M., Powell, A., & Maher, C. (2010). Expanding participation in problem solving in a diverse middle school mathematics classroom. *Mathematics Education Research Journal, 22*(1), 91–118.

Wei, M.-H., & Eisenhart, C. (2011). Why do Taiwanese children excel at math? *The Phi Delta Kappan, 93*(1), 74–76.

Zevenbergen, R., Dole, S., & Wright, R. J. (2004). *Teaching mathematics in primary schools*. New South Wales: Allen & Unwin.

Chapter 11
Commentary For Section 2: Inclusive Practices in Mathematics Teaching—The Need for Noticing and Producing Relevant Differences

Konrad Krainer

This book is a valuable effort to deal with a *grand challenge* facing our world, namely that our cultures become more and more heterogeneous, in particular through increasing worldwide immigration. This growing diversity has a deep impact on our society in general. Of course, this also influences our education system, and shapes new general conditions for teaching and learning in classrooms, at schools, and at teacher education institutions. Having students with different backgrounds in language, ethnicity, religion, parents' social status, etc., has to be taken into consideration seriously.

Whereas in the past *exclusivity* (and *differentiation*) was the dominant response to this situation, the authors plead for a clear focus on *inclusivity* (and *diversity*). I put differentiation and diversity in parentheses, since "differentiation" can be designed as *internal* or as *external* differentiation. The first could mean, for example, that the teacher(s) initiate(s) working groups within the classroom where students—due to their competencies and interest—choose different packages of tasks, however, exchanging experiences in a final classroom discussion; in contrast, external differentiation could mean, for example, to form three streams within one class through the whole school year, mirroring the idea of building continuous and heterogeneous sub-classrooms with different teachers. I assume that in this book differentiation is meant as external differentiation which indeed can be regarded as a stark contrast to an inclusive-bounded approach to diversity.

Consequently, the authors explore the implicit hypothesis that the greater the student diversity the greater is the need for inclusive educational approaches in mathematics education. At a first glance, it seems wise to design different approaches for different students, for example so-called "gifted" and "weak" students. However, this book opposes this labelling by arguing and developing other, more inclusive, practices. Thus, diversity is more regarded as a strength and opportunity than a weakness and thread.

K. Krainer (✉)
School of Education, Alpen-Adria-Universität Klagenfurt, Klagenfurt am Wörthersee, Austria
e-mail: konrad.krainer@aau.at

A. Bishop et al. (eds.), *Diversity in Mathematics Education*,
Mathematics Education Library, DOI 10.1007/978-3-319-05978-5_11,

This is a challenge, since the authors' view and the current situation at schools do not overlap very much. For example, Peter Sullivan highlights in Chapter 1 (2014, this volume) that a common tendency in schools (not restricted to Australian ones) is to offer "weaker" students a limited curriculum focusing solely on practical mathematics or low-level routines. This habit restricts these students' learning and career opportunities, and it also contributes to their alienation from school, and I would add from successful participation in society in general. In particular and above all these students would need learning opportunities where they are challenged to practise higher levels of mathematical thinking and action and to experience success. However, not having this experience, teaching increases the existing differences and reinforces the gap between "gifted" and "weak" students, with little or even no chance for the latter to develop adequately. There are many more such *visible* aspects (as the contrast between "gifted" and "weak") that make it so challenging to foster diversity and inclusivity. However, there are also manifold *latent* aspects that need consideration. Section 2 of this book focuses on these aspects.

This chapter has the task to comment on the previous chapters in Section 2 "Interrogating the boundaries". These three chapters scrutinise the latent aspects of mathematics education in relation to diversity. Latent aspects are not easy to grasp. Therefore, *expertise* and sensitivity are needed.

Here, it makes sense to indicate the difference between *experts* and *laymen*, as introduced in system theory by Willke (1999, p. 12; a link to mathematics education is done in Krainer 2005). Willke argues that the only feature that makes a difference between experts and laymen is experts' art of *precise observation*: an expert is able to see more, in particular differences, deviations, particularities, irregularities and peculiarities. He defines *observation* as *noticing* a relevant difference, and *intervention* as *producing* a relevant difference.

The relevant difference regarding this book is the contrast between *exclusivity* and *inclusivity*. The first represents—apart from innovative examples at some schools—the current situation, the second the desired target situation. The intervention (of the book) is to change the situation by indicating the need for new thinking about diversity research and practice in mathematics education and by proposing steps towards inclusivity.

A major group of stakeholders in this intended shift (one might even speak of a reform) are *teachers and teacher educators* (in the following I only refer to them as teachers). They need to understand the importance of this shift and to learn to notice relevant differences, and when noticed, eventually producing a relevant difference in their teaching. Thus, one goal of the book is to support teachers in becoming experts in noticing and producing differences with regard to diversity.

The first chapter focusing on latent aspects of mathematics education in relation to diversity is "Diversity, Inclusion and Equity in Mathematics Classrooms: From Individual Problems to Collective Possibility" by Mike Askew (2014, this volume). The chapter stresses that having a focus on the teacher or the learner at the centre of the diversity issue, the curriculum is often not questioned, thus mostly latent. Consequently, this chapter explores how current views of the curriculum influence the framing of diversity, and how alternative views might support inclusivity in mathematics classrooms and improve learning for all.

Askew argues that a more productive stance towards curriculum—than taking the individual as the starting point—focuses on creating *collective classroom cultures*, in particular on building *learning communities*. It is important to note that the focus on the collective is not understood as neglecting the needs of the individual, the support of the individual learner remains a major goal.

Referring to Smith and Barr (2008), Askew suggests thinking about curriculum as a *fact*, as an *activity*, and as an *inquiry*. The latter view takes a stance towards knowledge as co-constructed and emerging through interaction with others, thus shifting the focus away (from a transmission view and) from the individual learner and onto the activity of the collective. I follow Askew in stressing that curriculum as inquiry still has to be prepared for, even if lessons cannot be planned in the finest detail.

Referring to various authors, Askew introduces further differentiations, supporting the reader in noticing relevant differences. For example, he distinguishes between classrooms (1) as *communities*, (2) as *communities of learners*, and (3) as *learning communities*. Establishing classrooms as learning communities is regarded as a continuous and emergent process where structural and situational rationales for learning are important to take into account.

A strength of this chapter lies in providing manifold opportunities for *reflecting latent aspects* of diversity, meaningful links to a variety of other authors' theoretical considerations help to grasp the complexity. A further strength of the paper lies in making often *hidden assumptions* transparent, for example by acknowledging that diverse learners may be different from each other but not defining such differences in terms of deficiencies (Moschkovich et al. 2007). This helps reframing diversity away from being an obstacle to classroom talk being enriched through the diversity of learners' contributions.

The second chapter in this section is "Ethics and the Challenges for Inclusive Mathematics Teaching" by Helen Forgasz, Jennifer Bleazby, and Carly Sawatzki (2014, this volume). The chapter explores dilemmas and challenges of teaching mathematics when fostering *ethical understanding* which has been included as one of seven general capabilities teachers need to cover across all subject areas within the new Australian Curriculum.

The chapter starts with a vivid insight into each author's *impetus* for writing about ethics and mathematics learning, thus showing the diversity of situations where ethics is concerned and making some latent and personal aspects transparent and discussable to the reader.

The authors explore the *philosophical* and *psychological* bases of children's *moral development* in the educational context (e.g. referring to Dewey, Piaget, and Vygotsky). They argue that moral development is neither simply an internal, automatic process, nor a by-product of general intellectual development. Since successful moral development only occurs under particular conditions, teachers cannot assume that moral development occurs naturally without intervention. Moral development is dependent on teachers' noticing and producing differences with regard to diversity.

Like in the preceding chapter, also here the authors stress the importance of learning within a *community*. In particular, interactionist theories build on the belief that moral development is dependent on individuals interacting with their environments. According to Vygotsky's interactionist theory, learners need to be scaffolded by a more knowledgeable other, in particular by the teacher. Forgasz, Bleazby, and Sawatzki claim that the teacher should explain and model the more sophisticated moral inquiry skills that they want their students to develop.

The authors demonstrate how selected mathematical contexts and related social issues are value laden and ethically problematic. They successfully challenge the assumption that, unlike other subjects, mathematics is a value neutral discipline, stressing that an inclusive approach means that every facet of human experience and all knowledge domains have ethical dimensions. This is important for teachers to know, at the same time challenging, for example:

> "Presenting students with contextualised mathematical problems that raise obvious ethical issues, and then asking students to ignore the ethics and just focus on the mathematics, can send the rather dangerous message that, in mathematical contexts, ethical issues are secondary or can be disregarded altogether" (Forgaz et al., Chapter 9, p. 155).

The subchapter on the emergence of the *critical mathematics education movement* in the early 2000s is stimulating, in particular for those interested in further opportunities to deepen their knowledge. Very insightful are the four boxes with examples demonstrating opportunities to promote students' moral development, or inappropriate contexts which might cause the opposite.

In their conclusion, the authors highlight the fact that success in bringing these ideas into the classroom depends on teachers' willingness, knowledge, skills, motivations, confidence, and personal values and beliefs, and the sensitivity to poor and counter-productive examples.

The third chapter in this section is "Valuing Diversity in Mathematics Pedagogy Through the Volitional Nature and Alignment of Values" by Wee Tiong Seah and Annica Andersson (2014, this volume). The main focus of this chapter is on introducing a process called *value alignment*, and to illustrate how teachers can benefit from this approach.

As in the other chapters, the authors stress the *importance of teachers* when bringing about change. In particular, their capacity to have agency over what they value, and what they teach their students to value, is regarded as essential. In order to cope with the new challenge of increased diversity and to optimise mathematics learning, "teachers need to value the cultural diversity amongst their students and to value the diversity of mathematical ideas their students bring with them to class" (chapter abstract).

Seah and Andersson make a meaningful distinction between *two ways* in which values in relation to mathematics education can be considered: (1) teaching of values through the school subject of mathematics (*values through mathematics*) and (2) the facilitation of mathematics pedagogy through the harnessing of values (*mathematics through values*). The chapter focuses mainly on the second way, putting an emphasis on the valuing of diversity, both of students' ideas and reasoning, and of students' experiences and cultures.

A definitive strength of this chapter is the presentation and discussion of two scenarios from mathematics classroom teaching, using categories of values introduced by Bishop (1996), namely *mathematical* (the valuing of openness in the example), *mathematics educational* (fun), and *general educational* (respect). The same holds true regarding the examples used to highlight the significance of values in mathematics learning and teaching. This leads to thought-provoking insights, for example: "If teachers are not able to articulate their personal convictions, their values alignment experience to facilitate student learning will remain tacit" (Seah and Andersson 2014, this volume, p. 180). This means, again using the argumentation by Willke (1999): if teachers are not able to notice relevant differences concerning their values and thus are not able to articulate these, their ability to produce a relevant difference in teaching is limited.

Seah and Andersson stress that values alignment does not aim at a classroom situation in which most or even all people subscribe to the same values. In contrast, students perceive that their knowledge, skills, and dispositions are valued. Thus, the authors distance themselves clearly from *values inculcation*, their understanding is definitely a *(students') serious-taking* and *(self-)critical view of values alignment*. Putting an emphasis on valuing students' participation and perception, means that the same needs to hold true for *teacher education*, where (student) teachers' active and critical participation and perception is crucial.

All three papers in the Section 2 "Interrogating the Boundaries" are valuable contributions to inclusive practices in mathematics teaching and mathematics teacher education. It is important to promote teachers' expertise in noticing and producing relevant differences related to inclusive practices.

References

Bishop, A. J. (June 3–7 1996). *How should mathematics teaching in modern societies relate to cultural values—some preliminary questions*. Paper presented at the seventh Southeast Asian conference on Mathematics education, Hanoi, Vietnam.

Krainer, K. (2005). What is "Good" Mathematics teaching, and how can research inform practice and policy? Editorial. *Journal of Mathematics Teacher Education, 8,* 75–81.

Moschkovich, J. (2007). Bilingual mathematics learners: How views of language, bilingual learners, and mathematical communication affect instruction. In N. S. Nasir & P. Cobb (Eds.), *Improving access to mathematics: Diversity and equity in the classroom* (pp. 89–104). New York: Teachers College Press.

Smith, R., & Barr, S. (2008). Towards educational inclusion in a contested society: From critical analysis to creative action. *International Journal of Inclusive Education, 12*(4), 401–422.

Willke, H. (1999). *Systemtheorie II: Interventionstheorie*. Stuttgart: Lucius & Lucius UTB.

Part III
Towards Inclusive Practices

Chapter 12
(Dis)engagement and Exclusion in Mathematics Classrooms—Values, Labelling and Stereotyping

Alan J. Bishop and Penelope Kalogeropoulos

Introduction to Two 'Disengaged' Students

Tom is a grade 5 student attending a school in the south eastern suburbs of Melbourne. He was identified by his teacher as a disengaged student because 'as soon as your attention/focus away from them, they start talking or become off task very easily. They don't complete the work….' (VR 0:27–0:41). In class, Tom appeared to be easily distracted and unfocused. For example, in the first lesson, Tom placed his nametag on his mouth and pretended he could not talk. At another moment, he took a very long time to rule up his page.

Matthew is a grade 6 student who was chosen by his teacher and identified as disengaged. Matthew was a boisterous student, loud and frequently seeking attention from his teacher by constantly demanding assistance. When left to work independently, Matthew engaged in conversation with peers about matters other than schoolwork. Matthew arrived late to class (second lesson observed by researcher) because he was involved in a lunchtime incident and was reprimanded by the School Principal.

Introduction to This Chapter

> Student engagement is a highly complex and multi-faceted construct (Fielding-Wells and Makar 2008, p. 2).

> Researchers, psychologists and educators differ in opinions as to what constitutes engagement, how the construct can be measured and what aspects interact to result in engagement (Fielding-Wells and Makar 2008, p. 2).

A. J. Bishop (✉) · P. Kalogeropoulos
Faculty of Education, Monash University, Clayton, Australia
e-mail: Alan.Bishop@monash.edu

P. Kalogeropoulos
e-mail: penelopemonash@yahoo.com.au

A. Bishop et al. (eds.), *Diversity in Mathematics Education*,
Mathematics Education Library, DOI 10.1007/978-3-319-05978-5_12,

> Engagement is difficult to define operationally, but we know it when we see it, and we know it when it is missing (Newmann 1986, p. 242).

In the PISA 2009 results, Australia was ranked 15th for mathematics (down from 13th in 2006 and 11th in 2003). Student engagement is a topic of significant interest within Victoria and Australia in general, with a particular focus on student retention, participation and achievement (Zyngier 2008). The questions underlying this chapter are: Will these foci really address the problem of disengagement in mathematics in Australia? Why are our students increasingly underperforming in mathematics compared with relevant others? Is this a systemic problem or one that needs to be addressed at a more individual learner level?

This book's focus being on developing inclusive approaches to teaching and learning mathematics, we shall discuss what research can tell us that can help teachers deal inclusively with student disengagement. For example, exclusion practices often relate to the invisibility of the learners concerned, as often happens with gender and racial classroom mixes. (refer to Chapter 6 of this volume for a discussion on public mathematics stereotypes written by Gilah C. Leder and Helen J. Forgasz). In the mathematics classroom, however, disengagement is all too often visible, as can be seen with Matthew above. However, it can sometimes be the 'included' and 'engaged' students who are invisible to the teacher, with the 'quiet achiever' being a known phenomenon in mathematics education practices at all levels.

We will look at some of the existing and past research evidence on engagement and disengagement, and discuss issues such as the labelling of students and the problems that causes. We will also consider some of the related issues of what value-conflicts are implicated in the (dis)engagement process, and present data from a recent research study involving Tom, Matthew and other students. Finally, we shall speculate about possible new directions for research and for practice, and propose a new construct which may offer a more positive and inclusive approach to the (dis) engagement situation (note that this bracketing of 'dis' signifies that we are talking about engagement, disengagement and re-engagement together).

Research on (Dis)engagement

Most relevant research acknowledges that educational (dis)engagement is a complex phenomenon, consisting of diverse features. Typical of this research is that by Kong et al. (2003) who claim that engagement is best represented by the following three dimensions: affective, behavioural and cognitive, with which they developed a framework to measure engagement in mathematics. The study resulted in the identification of significant markers of engagement which included:

Affective engagement—interest, achievement orientation, anxiety and frustration

Behavioural engagement—attentiveness, diligence, time spent on task, non-assigned time on task

Cognitive engagement—surface strategies (memorisation, practising, test-taking strategies), deep strategies (understanding, summarising, making connections, justifying), reliance (on teachers/parents). (Kong et al. 2003, p. 5)

So what does this analysis enable us to recognise regarding disengagement? Can we just turn these ideas into aspects of disengagement, and argue that, for example, concerning the affective dimension, if a student is *not* showing any interest in achievement this would be a sign of disengagement? Similarly, regarding the behavioural dimension, if a student is *not* attentive or diligent, can this express disengagement? Also regarding the cognitive dimension, can disengagement be inferred if a student is failing to show understanding in mathematics? In short is disengagement just the opposite of engagement? Or is it the lack of engagement? If a student is not engaged is she disengaged?

It is likely, however, that in order to clarify disengagement effectively, one would need to identify markers in some combination of all three dimensions simultaneously. For example, it could be said that a student is not showing interest in class due to the teacher's presentation of the subject matter, hence a marker of disengagement. However, the student could be described as engaged because their performance level is adequate (e.g. from accurate assessment records).

(Dis)engagement and the Individual Learner

'Focusing on a lack of student engagement assumes that the problem is with the students; that is the students are in need of changing' (Zyngier 2008, p. 1766). However, the responsibility for the (dis)engagement issue cannot be left at the doorstep of the student. Teachers, parents and other significant adults must also accept responsibility. For instance, one significant problem with focusing on the individual level is that this leads to labelling the student as 'at risk, underachiever, slow learner, etc.'. Is this really a step towards resolving the problem? We will return to the problems with this practice later.

Clearly, there is individual-level engagement associated with a learner's personal valuing of mathematics, motivation, anxiety and future intent. For example, there has been extensive research into mathematics anxiety (Bessant 1995; Pajares and Urdan 1996), and it is noteworthy that disengagement was strongly predicted by this factor, while other research shows significant links between mathematics anxiety and avoidance of the subject. On the other hand, at the classroom level, students show episodes of engagement by successfully completing the tasks set by the teachers. In addition to this, it has also been noted that within the classroom context, 'the ability level of the group in which one resides can impact self-relevant academic processes and outcomes' (Martin et al. 2012, p. 5).

Vale and Bartholomew (2008) cited a finding from the PISA 2003 study relating to female students' confidence in mathematics, that is that compared to males, they appeared to be 'less engaged, more anxious, and less confident in mathematics' (p. 279). Overall, high-achieving boys appeared to be more confident than girls about mathematics. Moreover, as Barkatsas (2012) says 'Compared to their female peers, boys also demonstrated stronger behavioural and affective engagement, more confidence in using computers and CAS calculators, and had a more positive attitude to learning mathematics with computers and CAS' (p. 167). Please see Chapter 5 of this volume written by Hazel Tan who discusses the use of technology in mathematics education and gender bias.

Thus, it is clear that there is evidence showing the ways that intraindividual factors can impinge on (dis)engagement, but let us now turn to the student's educational and social contexts.

(Dis)engagement and the Ecosystem of the Student

Turning to some of the background research, which locates the student within the wider educational and social context, we ask: Does the wider context have an influence on (dis)engagement? If so, how and why? For example, in a large-scale study of student disengagement and the wider school and societal environment, Martin et al. (2012) showed that the variables of student, home and classroom explained most of the variance in the dependent variables. However, accounting for the disengagement variance specifically, the additional predictors were mathematics anxiety, perceived classroom engagement, the school's non-English speaking background (NESB) composition and the school's socio economic status (SES). Yair (2000) similarly argued that a multitude of factors (individual, gender, home, classroom, school) operate to impact students' engagement with the curriculum, while according to Hattie (2009), 'students not only bring to school prior achievement…but also a set of personal dispositions that can have a marked effect on the outcomes of schooling' (p. 40).

More generally, research has often indicated that home factors can significantly influence the educational development of a student by, for example, providing parental support in the form of affective engagement—e.g. showing interest towards one's child's learning, with empathy towards their feelings and frustrations. Furthermore, in the cognitive domain of engagement, providing children with resources (e.g., calculators, times tables charts) to support their learning can also achieve positive educational outcomes.

The home context can of course impact student learning in various ways. For example, immigrant children need to adjust to a new set of values in their new environment and run the risk of a conflict between home and school values. Students with a low SES, or coming from immigrant communities who have suffered from poverty and trauma, generally perform less well in mathematics than their peers. However, it is important to note that in Australia the educational systems do tend to provide increased funding to schools based on their ethnic composition. Where this is the case, there may be a positive effect on academic processes and outcomes of NESB students. It is also important to note that relatively well-established ethnic groups can have differential effects on (dis)engagement and achievement (Martin et al. 2011).

The size of the school can also influence the learning of the students. Small- and medium-sized schools seem to promote a more positive learning environment, as the students experience greater collaboration, share more decision making and have higher quality interpersonal relationships (Lee and Smith 1997; Newman et al. 2006). On the other hand, it is acknowledged that large schools may be better able to offer strong core curriculum and electives (Hattie 2009). A recent Australian political initiative has been to increase the teacher-to-student ratio so that there will be greater opportunity for individual attention to students (and better relationships formed with facilitating teachers), small group learning and more efficient processes to assist students' academic development. This initiative has led to improved student engagement and learning (Hedges et al. 1994). These findings also underscore the importance of the ecological systems approach to research into (dis)engagement. More recent research has explored the notion of reconfiguring teaching/learning into 'place-based pedagogies', as is discussed in Chapter 3 of this volume by Debra Panizzon. This implies that the cultural, political, economic and ecological dynamics of places should be discussed whenever we talk about the purpose and practice of learning (see, for example, Gruenewald 2008).

Clearly, the cumulative effect of these studies is to show us that the professional actions of teachers in a classroom can only do so much, and therefore much will depend on the values' and beliefs' context, as portrayed by the school system and the home, rather than just by individual classroom teachers and students. As will be seen in the section on implications for practice, it seems addressing re-engagement must also encompass school-level attention and more extensive student, home, classroom and school factors.

Values, Beliefs and (Dis)engagement

In considering the ways that (dis)engagement relates to other researched variables, the constructs of values and beliefs come to the fore, as has just been noted. For example, it is evident that there are values implicated in the list of markers given earlier by Kong et al. (2003). If a student values *diligence*, they probably would not be disengaged in class because they would want to maximise their performance in the subject. Other markers listed in the Kong et al. (2003) structure reveal the presence of other values, for example *reliance, interest, achievement orientation, anxiety* and *frustration*. For the markers *'Time spent on task', 'Non-assigned time on task', 'Surface strategies' and 'Deep strategies'*, value indicators could help to define and clarify these further within the classroom context.

Not only would research focused on values be helpful for the purposes of clarifying (dis)engagement but it would also help when considering explanations of student behaviour which might cause a teacher to inappropriately describe a student as either an engaged, or a disengaged, learner. Recognising that generic values and valuing might lie behind, and account for, students' specific behaviours not only bring another dimension of explanation, but it could also offer other pedagogical strategies for re-engaging students. Instead of focusing on the behaviours themselves, which might be thought of as merely the symptoms of (dis)engagement, teachers could begin to address what it is that the students are valuing, and which may need to be recognised, and perhaps problematised, if the complex process of re-engagement is to be fostered.

This is not to suggest that changing, or developing, students' values is an easy thing to do—far from it (see Bishop and Seah 2003). Experienced teachers who were interviewed about the values they were trying to teach were also unwilling to claim that they had succeeded in changing their students' values. The actual teaching, or intentional development, of values does continue to be a fraught area in research, and in practice. This point will be taken up again in the section on implications for practice.

However, related research studies have clarified that there is a close relationship between values and beliefs for teachers (Bishop et al. 2001). One possible description of the relationship is that values are preferred beliefs in action. That is to say that a teacher may hold several beliefs such as: 'all students can learn mathematics, all students have their preferred ways of learning', but it is when choices have to be made in action that their values (as preferred beliefs) are revealed. Indeed, it is often the case that some beliefs can be contradictory, for example 'all students can learn mathematics' versus 'not all students should learn mathematics throughout their schooling'. Dealing with those kinds of conflicts can reveal teachers' values, and perhaps the more cognitive idea of beliefs, and believing, is rather easier for teachers to address compared with the more affective variables of values and valuing. Moreover, some teachers in the study referred to above claimed that they did not teach any values when they were teaching mathematics. Indeed, one teacher went so far as to say that he enjoyed teaching mathematics precisely because he did not have to teach values!

Teachers and researchers have often noted that students' attitudes and beliefs affect their ability to explain, learn, and engage with classroom mathematics. Students' beliefs about the nature of mathematics, about its usefulness and about one's ability to do mathematics, all influence student achievement and values. For example, students who are not provided with opportunities to explore mathematics related to their daily lives may feel that mathematics is unimportant for them, and may become disengaged in their mathematics class, paying less attention and generally hampering their learning (Dossey et al. 2002). Thus, these researchers recommended that teachers should:

- Make mathematics meaningful;
- Motivate learning;
- Develop students' confidence in their ability to do mathematics, and,
- Help dispel myths relating to mathematics. (Dossey et al. 2002, p. 58)

The above list of recommendations is about the beliefs of students and their ability to do mathematics. However, given the relationship between beliefs and values (Bishop et al. 2006) it is clear that values are implicitly referred to also. In making mathematics meaningful, we value its importance in the world. In motivating learning, we are valuing the idea that mathematics is intrinsically interesting. In developing confidence in the ability to do mathematics we are valuing the fact that mathematical activities are observable and developable. In dispelling myths relating to mathematics, we are valuing *openness* (see Bishop 1988) in challenging and critiquing theoretical ideas.

But who is 'we' in the above paragraph? Teachers, students or both? Although teachers might aim to deliver mathematics in an interesting manner, as indicated above, student disengagement can continue to occur. Evidently, more research is required in relation to values, and we can, and should, ask to what extent are disengaged students' values supported by, or in conflict with, those behind the mainstream classroom practices?

Perhaps there is something more that we can do to understand disengagement, by looking beyond lists of affective, behavioural and cognitive markers. These markers appear to focus disengagement on the individual student: how one feels, how one acts and how one thinks. However, the individual operates in a sociocultural environment, and his/her interaction with the environment will likely regulate the affective, behavioural and cognitive processes. In this light, can we learn from the values' experiences of learners in cultural conflict situations?

For example, the composition of a multicultural classroom involves students who bring with them values, beliefs and attitudes that are influenced by their own home cultures. The schooling of immigrant children is a transition process such that when they arrive into a new country they are required to cope with the many changes involved in moving from one culture's values to another's (Gorgorió et al. 2002). These authors argue that it is important for teachers to:

- Acknowledge the student as an individual;
- Understand the meanings that students attach towards people and their environment;
- Acknowledge the student as a member of the classroom community;
- Recognise the meanings as social products developed from social interactions between members of the classroom community; and
- Acknowledge the student as an individual with a socio-cultural identity (see Gorgorió et al. 2002, p. 33).

Students attach meanings to situations, to actions, to themselves and to others through an interpretative process, which is revised and controlled through the acquisition of new experiences. 'Valorisation as a part of the interpreting process projects the sociocultural identity of the student within the mathematics classroom' (Gorgorió et al. 2002, p. 33). Therefore, our students bring with them different opinions as to what constitutes effective mathematics learning and classroom behaviours.

Learners who may be influenced by significant older others (e.g. their parents), may be led to believe that effective mathematics learning is about passing or excelling on a test, whereas their teachers and peers may feel that understanding the mathematical principles are most important. Thus, awareness of the roles of values in the classroom interaction process seems to be a significant factor in teachers' understanding of (dis)engagement. However, let us now turn to one common feature of teachers' practices which could well be having an effect on whether students engage or not with mathematics learning.

Labelling, Stereotyping and Significant Others

Research into affective issues in psychology and education has shown that a major cause of learning problems is the labelling of students by teachers and others, as, for example, when describing students as 'of low ability', 'a slow learner', 'a gifted student' or as in this case, 'disengaged'. Labelling a student is not just a simplistic and limited descriptive process, it can often have permanent and self-fulfilling effects for that student. In the area of special education, McDermott (1996, 1974) has shown the damaging effects of labelling, in the provocative message of his well-known paper: 'A disability in search of a student'. This occurs where a student's abilities are not seen in the richness which all students possess and deserve, but are limited to one closed and impoverished set of characteristics going by the disability.

Moreover, in relation to the theme of this book, labelling is a classic way of excluding learners. If you are called, or thought of as, a 'slow learner' for instance, then the teacher may well decide that you will need some special treatment, which on the face of it may seem like good practice. But it often means that you are thereby excluded from the mainstream teaching of the rest of the class. In addition if the teacher has low expectations of your achievements compared with your peers, this 'special treatment' will signal those low expectations to the rest of the class, and it will be difficult to avoid living up to, or rather down to, them yourself.

Mathematics in particular is a subject which encourages elites and elitism, while at the same time developing negative feelings and ultimately disengagement in others. Speed of solving problems is highly valued; speed in calculating also is valued. The highly abstract mathematics curriculum can be an interesting area to explore for those who find it easy, but it can also be a daunting obstacle for many learners. This can easily lead to disengagement, manifesting itself in ways that attract pejorative labelling of some students, such as 'slow learner', 'disadvantaged', 'weak', etc.

Labels are used as a shorthand for a range of characteristics, and it is but a simple step from labelling to stereotyping, and there is much research showing the negative influences of stereotyping, with gender stereotyping being the most researched (see, for example, Leder 1992). In this situation, the learner comes under the powerful influence of 'significant others'. In educational contexts, these persons are chiefly teachers and parents, and they are called significant precisely because they can exert a powerful psychological influence on the learner (Woelfel and Haller 1971).

However, what makes the classroom context such a powerful location for stereotyping is that it is often the students themselves who classify and categorise their classmates with various labels, and then, of course, they act in ways which reinforce them. Data from a research study show some good examples of such a problem—that of recent migrant arrivals into a new class in Australia (Bishop 2001). The following exchange is part of an interview with a recently arrived student from Europe, in the seventh-year mathematics class:

> Int: What's it like being in your class?
> Gor: I don't know, they call me a square because I know more than normal kids. I don't like them calling me a square.
> Int: Tell me a bit more about that. Who calls you a square?
> Gor: All the kids who don't know much.
> Int: Why do you think they do that?
> Gor: I don't know, because they are jealous.
> Int: How do you cope with that? Do you say anything back?
> Gor: I just ignore it.
> Int: Does it get worse if you ignore it?
> Gor: They are just like, sometimes I say 'thanks' and I like them saying it, so it sounds like I like them saying it, so they will probably stop saying it!

Also interesting is that student's use of the word 'normal'—another pejorative learnt label. Another recent migrant in another class uses a similar strategy:

> Int: Some students I have talked to have felt pressure from other classmates—when they do well they get called 'a square'.
> Mar: I don't worry about that, my life is sort of, like I don't live my life for those people, they are obviously not my friends if they say that. And if they are just joking I don't mind.
> Int: So do you feel their pressure at all?
> Mar: No, I probably would if I paid any attention to it, not really, my friends don't, maybe as a joke sometimes. I don't mind that.
> Int: So you decided just to ignore it?
> Mar: Yes, it's the marks at the end that is more important

So labelling per se is not necessarily a problem in this classroom context if the students concerned are able to stand up for themselves. Clearly, a less confident student than Gor or Mar would have a more difficult time in that kind of situation.

Another issue can occur when the negative labelling of erroneous actions or practices generalises to the negative labelling of the learner and their abilities. Actions and practices can be problematised and modified, but labelling the person and not the behaviours can have long-term negative consequences, and in educational contexts can certainly result in negative self-perceptions leading inevitably to learner disengagement. It is difficult to change someone else's perception of them, and how often do we hear of adults who feared their mathematics teacher who always described them as 'weak' or 'of low ability'?

In these two sections of the chapter, we have discussed values, beliefs, behaviours, labels and stereotypes, and how they relate to learner (dis)engagement. However, unlike in the section on research on (dis)engagement where we recognised that perhaps it is necessary to go beyond learner behaviours to try to understand more about the values that the disengaged student is pursuing, here we are proposing that a focus on the diversity of learner behaviours would help to get away from inappropriate labelling and stereotypes.

This contrast is less a conflict of ideas and more about what practices could be appropriate for encouraging engagement, or stimulating re-engagement, in the classroom. More discussion of the implications surrounding appropriate teaching practices will occur in a later section of this chapter.

In the next section, we will consider some data from an empirical study which concerns Matthew and Tom, two students who were both identified by their teacher as disengaged.

A Relevant Research Study—Into the Classroom

Relevant theoretical ideas from literature have been addressed in the previous section. What do these imply for the classroom situation? Ideas from this section will help us to develop implications for teaching practice. In particular, one of the benefits of qualitative research is being able to focus on the students as individuals so we can focus on specific implications for their inclusion and engagement. The next two sections of the chapter will develop these implications.

A recent study (Kalogeropoulos 2014), has shown that mathematics students labelled by their teachers as 'engaged' or 'disengaged' often show a mixture of engaged and disengaged practices. Thus, rather than using a labelling approach for tackling the disengaged students, another hypothesis is that it would be better for teachers to focus attention on the classroom learning and teaching conditions, which encourage student engagement or discourage disengagement.

The significant markers in the engagement framework proposed by Kong et al. (2003) will now be used to analyse episodes of engagement and disengagement by Tom and Matthew. These two boys were labelled as 'low achievers' in mathematics by their teacher. They were placed in a group of students with supposedly similar learning needs based on their prior performance in the nation-wide test known as National Assessment Program—Literacy and Numeracy (NAPLAN), which is an annual assessment for all students in Years 3, 5, 7 and 9. It tests the types of skills that are essential for every child to progress through school and life, in reading, writing, spelling, grammar and punctuation and numeracy. The assessments are undertaken in the whole of Australia, every year in the second full week in May.

With regard to the cognitive engagement dimension (surface strategy and reliance) by Kong et al. (2003), both Tom's and Matthew's academic performance in mathematics and demand for constant teacher assistance throughout the lesson demonstrated episodes of disengagement. However, during a mathematics game played in class, both students showed a level of sophisticated thinking by utilising mathematical strategies to improve their chances of winning the game. This could be classified as cognitive engagement (deep strategy).

Tom was easily distracted in most of the observed mathematics lessons. He was commonly seen to use the class materials as 'toys' and only persisted with the set independent learning tasks when he was supervised by his teacher. In regards to the behavioural engagement category, Tom exhibited disengagement by his inattentiveness, lack of diligence and his limited time spent on tasks that he was required to complete independently. In essence, Tom shows the value of *reliance* on his teacher.

What is of interest however, is Tom's behaviour during mathematics games. In these sessions, he showed contrasting behaviours. For example, he became interested and attentive to the assigned task. It could be suggested that Tom values *fun* in mathematics. Previous research has shown that the positive role of student-level enjoyment is important in addressing disengagement (Brown et al. 2007; McPhan et al. 2008; Nardi and Steward 2003). It is therefore worthy of consideration in educational practice more widely.

Matthew was labelled as a boisterous student in the observed mathematics lesson. He was commonly heard to 'call out' and interrupt the class and their teacher. He seemed to make an attempt on the assigned tasks but his strategies were sometimes inaccurate and repeated due to the lack of a 'check system'. This is an example of an engagement episode (his eagerness to complete the task) and a disengagement episode (inaccuracy overlooked) in the behavioural and cognitive domains of the framework by Kong et al. (2003). Matthew showed an element of affective engagement when he declared his frustration towards a teacher's request to repetitively write out the times tables as a form of discipline.

Having clarified some of the issues and background ideas, we will now present more specific aspects from the research with the two students.

Tom's Interviews and Classroom Episodes

In the first interview, Tom could not recall a mathematics lesson where he was not learning mathematics well. Also, he explained to the researchers that the best mathematics lesson he had ever done was 'word problems'. However, Tom appeared disinterested and off-task in the observed 'word problem' lessons. This contradicting information could suggest that Tom is simply referring to his recent mathematics lessons rather than focusing on what he enjoys in mathematics.

In lesson 2, the teacher allocated a great amount of time in helping Tom with his work. During this one-on-one teaching, Tom was attentive and focused on his work; however, as soon as the teacher moved away from his working area, Tom used the manipulative materials (Unifix blocks) to make models of guns. He clearly exhibited again his valuing of *reliance* on his teacher.

In the second interview, Tom explained that the noise level in his learning environment did not bother him because he was used to it. He also explained that he does not like demonstrating on the board because he is not good at writing on the board or at public speaking; thus the value of *confidence* is important here. Tom's teacher instructed him to move to a different seat and when Tom was asked by the researcher why his teacher made this request, he said that her intention was for him to see the board better. However, the teacher's intention was actually to stop Tom from distracting his peers.

In the interview Tom was able to nominate an engaged student in his class but not a disengaged student. He also spoke about the fact that he only shows 'good work' to his mother and does not usually discuss his class work with her or show her his workbooks. Is Tom's home context supporting him in his learning if he only feels confident to show a parent his 'good' work? Would Tom be more engaged with his learning if he was able to discuss and address his mathematical weaknesses with a parent?

In lesson 3, Tom used the Unifix blocks to make different models (such as towers and guns). He attempted to distract the student working next to him by inviting the peer to look at some pictures he had drawn. During the teacher demonstration, he made limited eye contact with the teacher and did not look at the teacher's demonstration on the whiteboard. During class time, he asked to get a drink of water but his request was declined.

When questioned as to why the teacher chose to play a game in class, Tom thought they had extra time left over and did not mention anything about student confusion in the lesson. Tom did not seem to be interacting actively in his classroom context. He seemed distant and excluded from the 'flow' of the lesson.

The picture in Fig. 12.1 shows Tom's drawing of an ideal mathematics classroom. It is very similar to his current classroom setup. Cooking, computers, whiteboard and multiplication tables charts are all visible. The speech bubbles exclaim 'Yay, Cool and Wow'. Tom explained that the people are saying 'Yay, Cool and Wow', '*because they got a new laptop computer*' (VR 0: 20:54–0:20:59). Tom refers to the setup of the classroom and the different experiences that are conducted in this environment. He refers to some mathematical experiences, for example the times table chart and the teacher-directed teaching with the child writing an answer on the board. When asked if cooking is part of mathematics learning, he answered 'kind of' (VR: 0:25:30 − 0:25:31). When asked about computers and mathematics learning, Tom referred to Mathletics (a mathematical computer program). Tom seems to have a passion for technology. Perhaps this is a learning tool best suited to Tom's mathematical learning?

Tom's answers to the student questionnaire (see Appendix 1) suggest that he prefers to learn mathematics through real world problems, investigations, mathematical games, models and materials, mathematics projects, posters and displays. His answers to questions 3 and 4 of the student questionnaire indicate that he understands the importance of mathematics and that his learning in this subject is a positive experience. It is interesting to note that Tom believes that he is a capable student in mathematics. This demonstrates his lack of self-efficacy and understanding. His teacher has labelled him as a disengaged student for this study, and during classroom observations and through interviews, Tom showed both cognitive and behavioural characteristics of disengagement.

Tom did not refer to any emotional aspects, nor did he show any real signs of emotional distress. Is this possibly why his teacher did not describe any emotional (dis)engagement characteristics of Tom when asked to explain the criteria she used to identify him as disengaged? Is Tom camouflaging his emotional distress or is he not aware of his real performance in mathematics? Is it possible that Tom's significant others have focused on positive encouragement and therefore led Tom to have a positive belief about his learning in mathematics? But how does this help this student to progress and develop his understanding of mathematics?

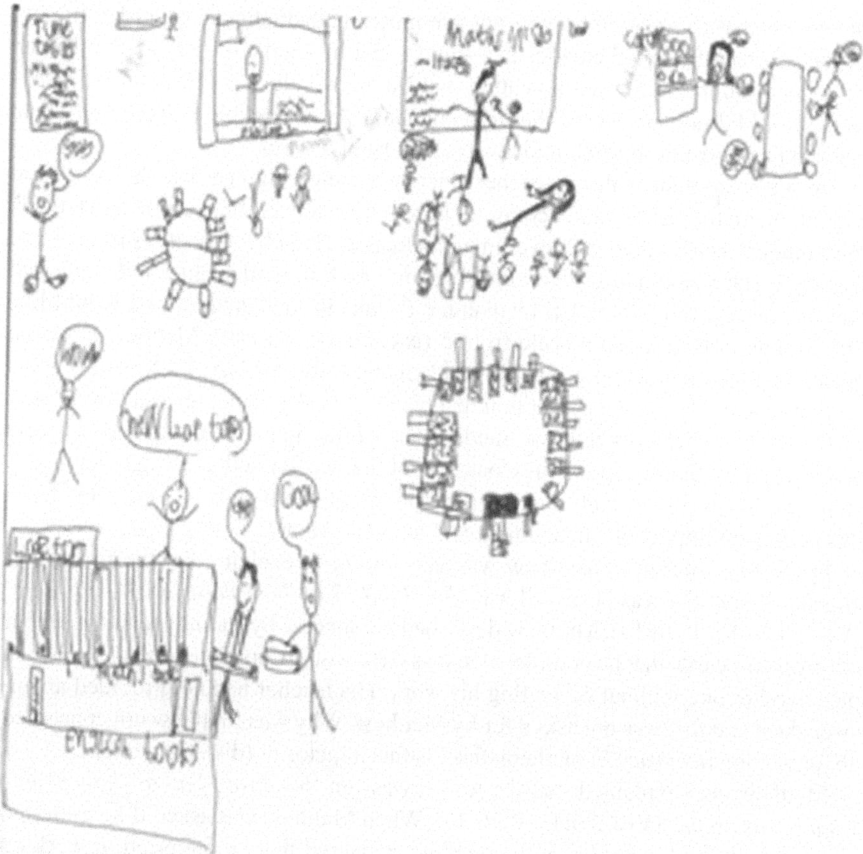

Fig. 12.1 Tom's classroom

Matthew's Interviews and Classroom Episodes

In the student questionnaire (see Appendix 1), Matthew answered that he always completes the activities listed in question 1. He answered that often the whole class talks about his work together, practises mathematics skills and explains his ideas to other students. Matthew answered that he sometimes talks about his work in small groups (valuing significant others) and completes investigations. In question 2, Matthew answered that he preferred to complete investigations rather than practise mathematics skills. He also prefers to do mathematics projects rather than solving problems.

For the remaining statements, he recorded that he equally preferred the listed activities. In question 3, Matthew said that he likes mathematics because we try to solve problems that we have and mathematics is important for his future because it helps him to think. In question 4, Matthew said that he strongly agreed with the statements that as an adult he will not use much mathematics in everyday life but strongly agreed that mathematics is a subject that he needs to study so that he can get a good job in the future, indicating that Matthew values mathematics as a subject. Also, Matthew answered that in mathematics it is possible to have more than one right answer but there should always be a right answer.

He seems content with his teacher since he strongly agreed that she is good at explaining mathematics and encourages him in this subject. In question 5, Matthew rates himself as a medium achiever in mathematics. He believes that his teacher and friends in class would also give him a medium rank in mathematics but suggested that his parents think he is below average (weak) in mathematics and would like him to rank only a '2' on a scale from 5 (excellent)–1 (weak). Matthew has been placed in a 'low achieving group' for mathematics after his NAPLAN results at his current school. He has the perception that his parents also rank him as a low achiever. However, he believes he has a 'medium' capability in mathematics and his peers would agree with him. In his previous school, he was ranked as a 'high achiever'. Here, we have lots of contradicting perceptions of Matthew's self-efficacy. Does this confusion impact on engagement and to what extent?

In the first interview, Matthew was very proud to exclaim that the first mathematics lesson that was observed was '*fun. I did 17 of those cards and I enjoyed it*' (VR1: 0:26:12–0:26:19). Matthew described his success by stating the total number of (problem) cards that he completed in class. He would complete one card and then pick another one without correcting his work. His teacher had not provided an answer sheet and this was not asked for by Matthew. Why was Matthew not concerned about valuing *accuracy* in mathematics? Is this a factor in (dis)engagement?

Matthew also explained that the lesson was fun '*because I had to show people what it was about*' (VR0:26:43–0:26:46). When Matthew was asked if he preferred to work with his friends or on his own, he answered that he preferred to work on his own. This was further supported by describing a situation where he was learning mathematics well. Matthew again referred to the observed lesson where he was required to complete word problems on his own. Interestingly, this was not noticed in the observed lessons. In fact, Matthew was constantly talking to his peers, asking for assistance, borrowing stationary items or stating the number of questions that he had completed. He described a learning situation when he was not learning mathematics well was when he had to write out the three times tables in mathematics as a punishment.

Matt: I did three and then I just sat around doing nothing
Int: Three what did you do?
Matt: 3 times tables and I didn't do the rest.
Int: What did you have to do with the times tables?
Matt: Write them down. Write some of them down.
Int: Why didn't you like that lesson?
Matt: Because I wasn't really interested in it that much.... The whole class did for punishment so I didn't like it and plus everyone was talking and distracting so I didn't do it. (VR1: 0:29:47–0:30:33)

Here we have another student who describes a punishment through a mathematics learning situation. When mathematics learning (in the form of repetition such as writing out times tables) is used as a disciplinary action, how does this affect student attitudes towards mathematics and how does this relate to (dis)engagement?

In contrast, Matthew referred to the lesson that was observed as his best learning situation ever and also that the worst mathematics learning situation he had ever had was the times tables situation that he had previously described.

In the next interview, Matthew explained that the second lesson that was observed was ‘fun and I enjoyed it’. He explained that he enjoyed the multiplication task cards. He asked for a calculator in class because he wanted to check his work but this was declined by the teacher. Is this a ‘conflict of values’ situation arising between teacher and student? Matthew completed his homework sheet the prior night and his mother had checked this work and confirmed it was all correct. He described the work at home as ‘easy’. He explained that his mother looks at his homework, not at his school workbook, but they do converse about ‘How is school going?’ At this point, Matthew is referring to a significant other and their influence on his mathematics learning. Is this one reason as to why Matthew is performing better in class than Tom? Matthew is aware of the fact that a parent will check his workbook and therefore may feel the need to complete work and document his learning in class.

Matthew was eager to say that he asked for a challenge because the class work that was set was too easy and he did not want to write out the answers because ‘it just hurts my hand’. He explained that at his previous school, he was ‘the smartest’ but at the current school, this was not the case. Matthew did not believe that his previous teacher (at another school) taught him anything and he described himself ‘Not as smart anymore. I don’t try as hard’. He also said that the mathematics lessons are too noisy and that this distracts him.

In the second interview, Matthew said that he was changing schools at the end of the week. When asked to describe what his classroom would be like if he was a mathematics teacher, this was his response:

> it’s like girl, boy seated but the friends have to go on separate tables and two on each table friends but girl boy row. We would have seats at the front like with those mini tables and we would have tables at the back.
> And what kind of activities would you do?
> Well, times tables like harder ones so it’s like a challenge and I will teach them like from starter ones, and then I’ll go medium then harder. (VR1: 1:53:50–1:54:33)

This description of Matthew’s ideal classroom represents a very traditional learning environment. Perhaps, Matthew is recognising the fact that he requires structure and direction with his learning; the value of *reliance* is again evident here.

Matthew was asked to draw a picture of the above and give it to his class teacher before he left the school but unfortunately this was not done. However, in a discussion with the researcher he described his ideal mathematics classroom. Matthew referred to times tables as a learning experience for ‘his students’ with an emphasis on scaffolding the understandings of these number facts through easy, medium and then difficult questions. The classroom setting sounds quite traditional with a strict seating arrangement that does not allow friends to sit together. Perhaps Matthew is beginning to explore the elements that distract his own learning? This again raises the issue of conflict of values with his teacher.

Implications for Research

The research project from which the extracts of Tom and Matthew's interviews and classroom experiences are taken was limited by time and circumstance to a PhD-sized study. Nevertheless one thing is clear: The data obtained have revealed a marked and significant disparity between the teachers' and the students' perceptions of their (dis)engagement.

For example, the students in the full study (16 overall) all accorded themselves a far more detailed and richer picture than their teachers did. Why is this? Is this a natural feature of any unbalanced power relationship such as that existing between teacher and student? Is it the case that the more pedagogical power the teacher has, the more their tendency is to simplify the teaching/learning situation? More power often comes with particular pedagogies: A 'chalk and talk' teaching strategy is likely to be accompanied by more teacher power compared with a more 'open' and 'discovery' pedagogical strategy where the power is shared.

Moreover, in terms of the categories of Kong et al. (2003), the teachers tended to focus their descriptions of the students on behavioural aspects and features, while the students were more aware of their own cognitive and affective features. Why would teachers do this? Is it natural for practical/procedural considerations to shape their professional perceptions and activities? It is of course easier to actually see and recognise behavioural markers rather than inferring affective and cognitive markers from students' responses to the classroom activities. Of course, students tend to think about themselves in rich ways but teachers must deal with the whole classroom group of disparate individuals. Hence, they often use the 'labelling' strategy, where the label covers, and thus simplifies, a multitude of behaviours and individual traits.

Perhaps the most surprising finding from the larger research study above is that the students seemed to be more aware of the diversity of classroom activities which affected their (dis)engagement than were their teachers. We can quote as an example the subtle knowledge of pedagogy that Tom and Matthew showed in their interviews. Also their experiences of being given a simplistic and routine task as a punishment strategy was revealing. They were fully aware that this was not an appropriate teaching method for them, and thus they felt excluded and further disengaged from the ongoing lesson. They knew that it was to control and discipline them rather than to help them to learn more mathematics. This example is not a criticism of the teachers in this study, but it does show that the classroom mix of students can necessitate the 'simplification' of the teaching situation, with pedagogical compromises everywhere. One can ask is it not the case that the labelling of perceived student disengagement helps the teacher to simplify and deal with one example of classroom complexity?

Other questions and research hypotheses arise from the study, and from the literature reviews given earlier. For example, there might be some interesting pedagogical moves developed from considering the issues concerning the relationship between students' (dis)engagement and their inclusion/exclusion in the classroom lessons. As we can see from the hypothetical table below the relationship between the two dichotomies is complex and probably significant:

A Inclusive and engaged	B Exclusive but engaged
C Inclusive but disengaged	D Exclusive and disengaged

For example, while achieving cell A seems initially to be the main target for effective teaching, and cell D to be the one to be avoided at all costs, we have also seen examples in the research study which show that cell C is not necessarily a productive goal for the teacher. The disengaged student may be included in the class's activities but increased engagement may not be automatically achieved. Cell B on the other hand could well be an effective goal for some students who may wish to get on with their own work, and be thoroughly engaged in it. Inclusion may in fact be a more achievable pedagogical goal than engagement, which suggests that moving a cell D student to cell A may be achieved more easily if the teaching proceeds via cell B rather than via cell C. The teacher could see 'moving' a student from cell B to cell A as a much easier pedagogical task than moving a student from cell C to cell A.

But who is doing the inclusion and exclusion—the teacher or the student? It appears from the above analysis that it is the teacher and his or her pedagogies. However, we could argue that a student who appears to be disengaged from a mainstream classroom activity is actually choosing to 'self-exclude' him or herself. Equally one who appears to be engaged is choosing to 'self-include' him or herself. It is the student who makes these choices not the teacher. All the teacher can do is to set up what might be considered as the appropriate classroom conditions and help the student clarify the values underlying their choice of behaviours.

Having explored some potential issues for further research into the relationship between (dis)engagement and inclusive practices, let us now turn to some of the practical implications of this research.

Implications for Practice

The data from the above study suggest that engagement prevails with what the teacher sees as effective teaching practices and principles. For example, Matthew responded positively to mathematical lessons that he felt were addressing his learning needs and that were interactive and interesting. This is a positive indicator for dealing with issues of (dis)engagement as it removes the focus from individual students and provokes instead an emphasis on the mathematical pedagogy and classroom context. For a long time, mathematics teaching and learning has focused on the value of control (Bishop 1988). In Australia, endless worksheets with marked correct and incorrect answers had prevailed in classrooms. In contrast, this study showed clearly that episodes of disengagement decreased when classroom practice followed an inquiry-based approach. Moreover, when tasks were set that involved group work accompanied by the requirement for high-level thinking, the students exhibited frequent episodes of engagement.

'Only when children learn what they want to learn and begin to take the responsibility for learning and living can they stay truly engaged' (Zhao 2012, p. 171). Zhao (2012) recommends that children will be more engaged with their learning if they are given the freedom to do what they want as this will enable them to discover and pursue their strengths, learn to take initiatives (become risk-takers), preserve exceptional talents and stay committed to their learning.

But if the issue of inclusion/exclusion seems to turn on just the classroom pedagogy, then we need to recognise that there are more significant and relevant issues of curriculum choices and whole school planning. This in turn focuses attention on the wider educational and social context. As an example of what is needed, a school in the south-eastern suburbs of Melbourne has developed a position document that describes the school's vision and mission. Selected theorists and principles of learning are referred to in this document and an explanation of how these pedagogies can be embraced is described through pedagogical practices, organisational structures, physical environment set-ups, assessment procedures and school leadership support.

In essence, the school has developed this document collaboratively (with staff, students and parents), to create learning environments 'where learning is maximised and school is a place of optimism, excitement and challenge. Students and teachers see each day as a journey, full of purpose where intellectual engagement and connectedness to the outside world are priorities' (http://www.woorannaparkps.com.au/wp-content/uploads/2012/07/Raison-detre.pdf, Accessed 13 Dec 2013). This is achieved through the planning of authentic, inter-disciplinary, research-based project tasks with teachers perceived as facilitators of learning and students are empowered to take responsibility for their learning. As effective as these pedagogical principles are in this case, we should in general avoid labelling schools and refer to 'schooling effects' rather than 'school effects' (Martin et al. 2011).

What is needed to help inclusion and eradicate disengagement, is firstly to recognise that teachers need positive markers and descriptors which reflect effective practices rather than the negative practices of labelling and other excluding pedagogies. Other research has provided helpful strategies based on principled practice, such as mathematics learning takes place when students' learning styles and emotional and multiple intelligences are addressed (Davidson et al. 1994; Gardner and Viens 1990).

Another example is mathematical tasks which are 'rich', i.e. open-ended leading to inquiry and research (Fielding-Wells and Makar 2008), seem much more likely to lead to engagement rather than disengagement. Assessment that is documented in a variety of manners can also capture students' thinking and creativity. Students need to see the relevance of their mathematical learning tasks so that they are meaningful to their lives. Martin (2007) and Balfanz et al. (2007), recommend that the relevance and importance of mathematics in students' lives should be made explicit during the teaching and learning of the subject. This could then also lead to increased engagement when the students themselves recognise the importance of mathematics and value it in their lives.

Useful as these suggestions and ideas are, they are too often situated in isolation from each other. This leaves the teacher feeling confused and uncertain as to how to proceed effectively and in a structured way. What are therefore also needed are ways of developing inclusive practices for use throughout a student's passage in school, and not just as a one-off isolated activity. Just as we sequence curricular ideas into a sensible whole throughout schooling, so there is a need to do the same with pedagogical practices. A mere mixture of interesting pedagogical activities may be inclusive and engaging for a while, but unless there is a general structure to the pedagogy it will not succeed in the long term.

One example of developing markers of an inclusive pedagogy is a framework that has been devised which offers teachers and schools ways of focusing inclusive pedagogies for particular students, within an overall inclusive pedagogical structure. Based on earlier work on the teaching of values (Bishop 1988) it is called the Mathematical Well-Being (MWB) framework (Bishop 2012; and see Appendix 2). It is a theoretical construct built on the strengths of the 'well-being' construct from medical education, plus the power of the highly successful taxonomic approach first developed by Benjamin Bloom and his co-workers (see Bloom et al. 1956). It has its roots in the affective domain described by Bloom, and re-emphasised by Kong et al. (2003). It is a marker scheme which structures the learner markers into an inclusive pedagogical sequence.

The particular strength of such a framework is that it helps teachers to recognise the longer-term significance of individual student markers for determining overall pedagogical practices. Just as a curriculum framework enables teachers to see how any particular mathematical activity fits into a structured whole, so the MWB framework allows teachers to see from individual student markers, how any particular pedagogical practice could fit into a structured whole. The MWB is in the process of development but it stands as an example of what is needed to assist teachers grappling with the issues of inclusion and (dis)engagement.

Appendix 1

STUDENT QUESTIONNAIRE (PRE LESSON)

Name______________________________

Class______________________________

School______________________________

Date______________________________

Dear student,

There are no correct answers to the questions in this questionnaire. Every student is different and we are interested in your personal views. No-one else but the researchers will see your answers.

Question 1

How often do you do these activities in your mathematics lessons?
Tick one box in each row

		Always in every lesson	Often in some lessons	Some- times in few lessons	Rarely or not at all
(1a)	Talk about your work in small groups				
(1b)	The whole class talks about your work together				
(1c)	Do mathematics problems in the real world				
(1d)	Use models and materials				
(1e)	Practise mathematics skills				
(1f)	Solve problems				
(1g)	Do investigations				
(1h)	Do mathematics projects				
(1i)	Explain your ideas to other students				
(1j)	Make posters and displays				
(1k)	Play mathematical games				
(1l)	Explore mathematical puzzles				

Question 2

Mark an X on the line to show how much you prefer one activity over another one at the other end of the line.

Talk about your work in small groups	————	Do problems in the real world
Practise mathematics skills	————	Do investigations
Explaining to other students	————	Play mathematical games
Talk about your work to the whole class	————	Use models and materials
Solve projects	————	Do mathematics problems
Make posters and mathematical displays	————	Explore puzzles

Question 3

Please write a number in for each statement ('1' indicates your first choice, '2' indicates your second choice, '3' your third choice, etc. down to '6' your last choice).

I like mathematics because….

(3a)	We get to discuss with each other
(3b)	We do lots of practical work
(3c)	We try to solve problems we have
(3d)	We get to discover new ideas
(3e)	We get to show the other how we do things
(3f)	We learn about important mathematical ideas

Mathematics is important for my future because:

(3g)	It helps me to think
(3h)	It is about solving problems I have
(3i)	It teaches me lots of useful things
(3j)	It helps me to be creative
(3k)	I learn to tell other about my ideas
(3l)	It shows me that all kinds of problems are interesting

Question 4

Rate 5 (Strongly agree)–1 (Strongly disagree) (Tick one box in each row)

		5	4	3	2	1
(4a)	Mathematics is one of the most worthwhile and necessary subjects to study at school					
(4b)	I am no good at mathematics					
(4c)	In mathematics class, I listen carefully and pay attention					
(4d)	Girls often have to work harder than boys to do well in mathematics					
(4e)	I get confused and frustrated when I do mathematics					
(4f)	As an adult I will not use much mathematics in everyday life					
(4g)	I study mathematics because I know how useful it is					
(4h)	Mathematics is a subject I need to study so I can get a good job in the future					

		5	4	3	2	1
(4i)	I give up trying to work on mathematics when I cannot understand it					
(4j)	Mathematics problems should always be solved by following rules					
(4k)	To learn mathematics you do not need to explain what you are doing					
(4l)	In mathematics, it is possible to have more than one right answer					
(4m)	In mathematics, there should always be one right answer					
(4n)	My teacher is good at explaining mathematics					
(4o)	During mathematics, we usually work on our own					
(4p)	Mathematics is like a different language to me					
(4q)	We use lots of materials (resources to learn mathematics)					
(4r)	My mathematics teacher thinks some problems are too difficult for me					
(4s)	My teacher encourages me in mathematics					

Question 5

Rate 5 (Excellent)–1 (Weak) (Tick one box in each row)

		5	4	3	2	1
(5a)	How good are you at mathematics?					
(5b)	How good would you like to be at mathematics?					
(5c)	Where would your teacher put you on this scale?					
(5d)	Where would your mother put you on this scale?					
5e)	Where would your father put you on this scale?					
(5f)	How good do you think your mother would like you to be at mathematics?					
(5g)	How good do you think your father would like you to be at mathematics?					
(5h)	Where would your friends in class put you on this scale?					

Appendix 2

Stages of Mathematical Well-Being

Stage 0: Awareness of mathematical activity At this first stage the learner is aware of mathematics, not as a coherent body of knowledge but as a collection of mathematical activities. There is an awareness of the different nature of these from other school activities
Stage 1: Recognition and acceptance of mathematical activity The learner recognises mathematics as a coherent activity, different from a language or a sport activity and it is accepted as a similarly worthwhile pursuit. The learner feels comfortable in the mathematical learning context, although having a passive acceptance of such experiences and being disinclined to seek them out

Stage 2: Positively responding to mathematical activity
At this stage, mathematical activity invokes a positive response. More than just acceptance of the activity, here there is a welcoming of it and some pleasure in its pursuit and in its achievement. This pleasure develops feelings of self-confidence and positive self-esteem, which reinforce the acceptance and worthwhileness of mathematical activity in general

Stage 3: Valuing mathematical activity
At this stage the learner appreciates and enjoys mathematical activity to the extent that there is an active seeking out of those activities, and of people with whom those activities can be shared. The learner reaches acceptably high (to them) levels of mathematical competence

Stage 4: Having an integrated and conscious value structure for mathematics
At this stage the learner has developed an appreciation of mathematics, of how and why they value it, and where that valuing might lead them in the future. An awareness grows of the human development of mathematical knowledge, and of one's place in the mathematical scheme of things

Stage 5: Independently competent and confident in mathematical activity
At this stage the learner is a fully independent actor on the mathematical stage. Sufficiently independent to be able to hold one's own in mathematical arguments at various levels, the learner is able to criticise other's arguments from well-rehearsed criteria

References

Balfanz, R., Herzog, L., & Mac Iver, D. J. (2007). Preventing student disengagement and keeping students on the engagement path in urban middle-school grades: Early identification and effective interventions. *Educational Psychologist, 42,* 223–235.

Barkatsas, A. N. (2012). Students' attitudes, engagement and confidence in mathematics and statistics learning: ICT, gender and equity dimensions. In H. Forgasz & F. Rivera (Eds.), *Towards equity in mathematics educating: Gender, culture and diversity* (p. 167). Berlin: Springer.

Bessant, K. C. (1995). Factors associated with types of mathematics anxiety in college students. *Journal for Research in Mathematics Education, 26,* 327–345.

Bishop, A. J. (1988). *Mathematical enculturation: A cultural perspective on mathematics education*. Dordrecht: Springer.

Bishop, A. J. (2001). The transition experience of immigrant secondary school students: dilemmas and decisions. In G. de Abreu, A. J. Bishop, & N. C. Presmeg (Eds.), *Transitions between contexts of mathematical practices* (pp. 53–79). Dordrecht: Kluwer.

Bishop, A. J. (2012). From culture to well-being: A partial story of values in mathematics education. *ZDM Mathematics Education, 44*(1), 3–8.

Bishop, A. J., FitzSimons, G. E., Seah, W.-T. & Clarkson, P. (12–17 July 2001). Do teachers implement their intended values in mathematics classrooms? *Proceedings of the 25th Conference of the International Group for the Psychology of Mathematics Education. Volume 2.* The 25th Conference of the International Group for the Psychology of Mathematics Education (pp. 169–176). Utrecht: The Netherlands.

Bishop, A., Gunstone, D., Clarke, B. & Corrigan, D. (2006). Values in mathematics and science education: researchers' and teachers' views on the similarities and differences. *For the Learning of Mathematics, 26*(1), 7–11.

Bishop, A. J., & Seah, W. T. (2003). Values in mathematics teaching—the hidden persuaders? In A. J. Bishop, M. A. Clements, C. Keitel, J. Kilpatrick, & F. K. S. Leung (Eds.), *Second international handbook of mathematics education* (pp. 717–765). Dordrecht: Kluwer Academic.

Bloom, B. S, Engelhart, M. D., Furst, E. J., Hill, W. H., & Krathwohl, D. R. (1956). *Taxonomy of educational objectives; Handbook I: Cognitive domain.* New York: Longmans.

Brown, M., Brown, P., & Bibby, T. (2007). *"I would rather die". Attitudes of 16-year-olds towards their future participation in mathematics*. In D. Kuchemann (Ed). Paper presented at the Proceedings of the British Society for Research into Learning Mathematics. London, England: British Society for Research into Learning Mathematics.

Davidson, J., Deuser, R., & Sternberg, R. (1994). The role of metacognition in problem solving. In J. Metcalfe & A. Shimamura (Eds.), *Metacognition*. Cambridge: MIT Press.

Dossey, J. A., McCrone, S., Giordano, F. R., & Weir, M. (2002). *Mathematics methods and modeling for today's mathematics classroom: A comtemporary approach to teaching grades 7–12*. Pacific Grove, CA: Brooks/Cole.

Fielding-Wells, J., & Makar, K. (30 Nov–4 Dec 2008). *Student (dis)engagement in mathematics*. Paper presented at the Annual Conference of the Australian Association for Research in Education (AARE), Brisbane, Australia.

Gardner, H., & Viens, J. (1990). Multiple intelligences and styles: Partners in effective education. *The Clearinghouse Bulletin, 4*(2), 4–5.

Gorgorió, N., Planas, N., & Vilella, X. (2002). Immigrant children learning mathematics in mainstream schools. In G. de Abreu, A. Bishop, & N. Presmeg (Eds.), *Transitions between contexts of mathematical practice* (pp. 23–52). Dordrecht: Kluwer.

Gruenewald, D. (2008). The best of both worlds: A critical pedagogy of place. *Environmental Education Research, 14*(3), 308–324.

Hattie, J. (2009). *Visible learning*. Oxford: Routledge.

Hedges, L. V., Laine, R. D., & Greenwald, R. (1994). An exchange: Part I. Does money matter? A meta-analysis of studies of the effects of differential school inputs on student outcomes. *Educational Researcher, 23,* 5–14.

Kalogeropoulos, P. (2014). *Switching off mathematics: Sociocultural values that inhibit learning.* (Unpublished).

Kong, Q. P., Wong, N. Y., & Lam, C. C. (2003). Student engagement in mathematics: Development of instrument and validation of construct. *Mathematics Education Research Journal, 15*(1), 4–21.

Leder, G. C. (1992). Mathematics and gender: Changing perspectives. In D. A. Grouws (Ed.), *Handbook of research on mathematics teaching and learning* (pp. 597–622). New York: Macmillan.

Lee, V. E., & Smith, J. B. (1997). High school size: Which works best and for whom? *Educational Evaluation and Policy Analysis, 19,* 205–227.

Martin, A. J. (2007). Examining a multidimensional model of student motivation and engagement using a construct validation approach. *British Journal of Educational Psychology, 77,* 413–440.

Martin, A. J., Bobis, J., Anderson, J., Way, J., & Vellar, R. (2011). Patterns of multilevel variance in psycho-educational phenomena: Exploring motivation, engagement, climate, teacher, and achievement factors. *Zeitschrift fur Padagogische Psychologie/German Journal of Educational Psychology, 25,* 49–61.

Martin, A. J., Anderson, J., Bobis, J., Way, J., & Vellar, R. (2012). Switching on and switching off in mathematics: An ecological study of future intent and disengagement among middle school students. *Journal of Education Psychology, 104*(1), 1–18.

McDermott, R. P. (1974). Achieving school failure: An anthropological approach to illiteracy and social stratification. In G. D. Spindler (Ed.), *Education and cultural process: Towards an anthropology of education* (pp. 82–118). New York: Holt, Rinehart and Winston.

McDermott, R. P. (1996). The acquisition of a child by a learning disability. In S. Chaiklin & J. Lave (Eds.), *Understanding practice: Perspectives on activity and context* (pp. 269–305). Cambridge: Cambridge University Press.

McPhan, G., Morony, W., Pegg, J., Cooksey, R., & Lynch, T. (2008). *Maths? Why not?* Canberra: Department of Education, Employment and Workplace Relations.

Nardi, E., & Steward, S. (2003). Is mathematics T. I. R. E.D? A profile of quiet disaffection in the secondary mathematics classroom. *British Educational Research Journal, 29,* 345–366.

Newman, M., Garrett, Z., Elbourne, D., Bradley, S., Noden, P., Taylor, J., & West, A. (2006). Does secondary school size make a difference? A systematic review. *Educational Research Review, 1,* 41–60.

Newmann, F. M. (1986). Priorities for the future: Towards a common agenda. *Social Education, 50,* 240–250.

Pajares, F., & Urdan, T. C. (1996). Exploratory factor analysis of the Mathematics Anxiety Scale. *Measurement and Evaluation in Counseling and Development, 29,* 35–47.

Vale, C., & Bartholomew, H. (2008). Gender and mathematics. In H. Forgasz, A. Barkatsas, A. Bishop, B. Clarke, S. Keast, W.-T. Seah, & P. Sullivan (Eds.), *Research in mathematics education in Australasia 2004 − 7* (pp. 271 − 290). Rotterdam: Sense Publishers.

Woelfel, J., & Haller, A. (1971). Significant others: The self-reflexive act and the attitude formation process. *American Sociological Review, 36*(1), 74–87.

Yair, G. (2000). Educational battlefields in America: The tug of war over students' engagement with instruction. *Sociology of Education*, *73*, 247–269.

Zhao, Y. (2012). *World class learners (Educating Creative and Entrepreneurial Students)* Publishers: Originally published in 2012 by Corwin. Republished in Australia by Hawker Brownlow Education.

Zyngier, D. (2008). (Re)conceptualising student engagement: Doing education not doing time. *Teaching and Teacher Education, 24*(7), 1765–1776.

Chapter 13
Capturing Diversity in the Classroom: Uncovering Patterns of Difficulty with Simple Addition

Sarah Hopkins and Celéste de Villiers

Number fluency is an essential component of building mathematical competencies. Number fluency refers to using developmentally appropriate skills that allow a person to apply facts and meaningfully learned procedures, flexibly and efficiently, to solve new and familiar number tasks (Australian Curriculum, Assessment and Reporting Authority (ACARA) 2010). Number fluency has been referred to elsewhere as adaptive expertise (Baroody and Dowker 2003) and is encompassed in the idea of having good number sense (McIntosh et al. 1992). As number fluency is related to cognitive development, what constitutes number fluency differs from one year to the next as children progress through the curriculum. In the early years of school (grades 1–3), number fluency includes developing efficient and accurate strategies for performing simple addition.

A recently published study has suggested that it is not just a few, but *many* children who do not meet curriculum expectations of fluency with simple addition (Cowan et al. 2011). We have been interested in how children perform simple addition for a long time. Our own experiences of being a teacher in secondary schools (first author) and primary schools (second author) had already alerted us to the difficulties children have with simple addition. The study by Cowan et al. has spurred us on in our efforts to try and capture what we had experienced in our own classrooms and to draw attention to what we think is a potential barrier to learning mathematics that is often overlooked.

We believe that the significant number of children who are not meeting curriculum expectations is a serious concern for educators. The efficiency and accuracy with which simple addition is performed is closely aligned with how well children understand key number concepts like the inverse relationship between addition and subtraction (Canobi 2009; Gilmore and Papadatou-Pastou 2009) and the property

S. Hopkins (✉)
Faculty of Education, Monash University, Berwick, Australia
e-mail: s.hopkins@monash.edu

C. de Villiers
Graduate School of Education, University of Western Australia, Crawley, Australia
e-mail: celeste.devilliers@ststephens.wa.edu.au

A. Bishop et al. (eds.), *Diversity in Mathematics Education*,
Mathematics Education Library, DOI 10.1007/978-3-319-05978-5_13,

of commutativity (Canobi 2009). Based on an iterative view of mathematical development (Schnieder and Stern 2010), poor fluency with simple addition will act as a barrier to developing key conceptual knowledge as emergent understandings of number are not reinforced by efficient procedures and attentional resources are not made available during performance to discern underlying number concepts. Poor fluency with simple addition is likely to restrict student's opportunities to learn mathematics.

We acknowledge that there is no shortage of studies that have investigated simple addition performance. It is well documented that children identified as having a mathematics disability or difficulty are less likely than their typically achieving peers to retrieve simple addition facts (e.g. Bull and Johnston 1997; Geary 2010; Geary et al. 2000; Jordan et al. 2003; Ostad 1997; Torbeyns et al. 2004). This research, however, has led to a deficit view of learning difficulties (e.g. Landerl et al. 2004; Swanson and Jerman 2006), suggesting that some children have a circumscribed deficit associated with fact retrieval (Geary 2010, Jordan and Oettinger-Montani 1997; Robinson et al. 2002). A deficit view promotes the knowledge and research of cognitive psychologists and neuroscientists who are keen to investigate the possibility of a congenital difference between children considered to be typically achieving and low achieving in mathematics. We wanted to investigate simple addition performance in a way that was connected with the work and knowledge of teachers. We chose to do this by capturing some of the diversity teachers contend with each day in their classroom and focusing on how one school's cohort of year 3 children performed simple addition. We believe that all children are capable of improving their fluency with simple addition and that it is the work of teachers who will make the difference. We start to unpack how in this chapter.

In the first section of this chapter we define what it means to be proficient with simple addition and describe how and when proficiency with simple addition is expected to develop. In the second section, we report findings from three studies revealing striking diversity in how simple addition is performed by year 3 children attending the same school. In the third and final section, we summarise the findings and highlight patterns of difficulty that emerged from these studies. We discuss how these patterns of difficulty are useful for developing inclusive practices where teaching is targeted to address children's specific learning needs.

Proficiency with Simple Addition

Simple addition refers to the process of adding together single digit numbers. Proficiency with simple addition refers to the accurate use of *retrieval* and *decomposition strategies* to perform simple addition (Cowan et al. 2011). Retrieval involves directly retrieving an answer from a network of associations stored in long-term memory (Ashcraft 1995) and is often called recall in curriculum documents. Decomposition strategies involve partitioning numbers to make use of retrieved facts (Siegler 1987); these have also been referred to as thinking strategies (Rathmell

1978). Decomposition strategies include strategies that make use of add-to-ten facts (e.g. $3+8=3+7+1$) and making use of tie facts (e.g. $4+3=3+3+1$).

How Proficiency Develops

Before developing proficiency, children can use a range of counting strategies to perform simple addition (Hopkins and Lawson 2002). These might include the *counting-all strategy* where the count is started at one, a *counting-on-from-first strategy* where the second addend is counted on the first addend and the *counting-on-from-larger strategy* where the smaller addend is counted on the larger addend (Carpenter and Moser 1984). The counting-on from larger strategy is the most efficient counting strategy as it requires the minimum number of counts—it is also referred to as *min counting (or the min-counting strategy)* in more recent research (e.g. Fuchs et al. 2010; Geary 2010).

Although children can be taught more efficient strategies such as min counting (Fuchs et al. 2010, Tournaki 2003), decomposition strategies (Steinberg, 1985) and retrieval (Fuchs et al. 2006; Poncy et al. 2007), many children construct and adopt these strategies for themselves when they are ready to do so (Carpenter and Moser 1984; Cummings 1988; Siegler and Jenkins 1989). Christensen and Cooper (1991) found that typically achieving grade 2 children benefitted equally well or better from practice with the strategies they used spontaneously, than from direct instruction in the use of more efficient strategies (including min-counting and decomposition strategies). Two prominent models in the literature are used to explain how children develop more efficient strategies for themselves, the strategy choice model and the schema-based model.

The strategy choice model (Shrager and Siegler 1998; Siegler and Shipley 1995; Siegler and Shrager 1984) emphasises the critical role correct practice has on learning to retrieve facts. This model explains how practice with *a backup strategy* (any strategy other than direct retrieval) leads to the strengthening of an association between a problem and its answer in memory, thereby making direct retrieval more likely in the future. Later versions of the model account for two additional features of development: (1) practice using direct retrieval increases the likelihood of direct retrieval in the future and (2) as backup strategies become more practiced, attentional resources are deployed to construct more efficient backup strategies. The strategy choice model predicts that practice with backup strategies will lead to an increase in the use of retrieval but only if practice results in correct answers.

The schema-based model (Baroody 1994; Baroody and Tiilikainen 2003) emphasises the critical role conceptual knowledge has in facilitating the development of more efficient strategies. The schema-based view emphasises how strategies used to perform simple addition are transformed as rules that embody number relationships become more automated and integrated into a person's mental representations of basic number combinations. Children's discoveries of patterns allow them to devise rules and utilise their existing knowledge so that facts are not learned in isolation from each other.

These two models are not necessarily in conflict with each other but can be viewed as emphasising different factors that account for the development of simple addition proficiency—one emphasises the role of practice with spontaneously used strategies and the other, the role of conceptual knowledge.

When Proficiency Develops

Curriculum documents provide a guide as to when proficiency with simple addition is expected to develop. In this chapter, content descriptions from the Number and Algebra strand of the Australian Curriculum (ACARA 2010) are used as a guide to where teaching is focused in years 1–3. Although the descriptions provided are specific to the Australian context, there are strong similarities between the stages described below and those found in other curriculum documents and standards (e.g. Department for Education 2013; National Council of Teachers of Mathematics 2000).

Around year 1, number fluency includes being able to solve simple addition problems using a range of strategies including counting-on strategies (i.e. the counting-on-from-first strategy and/or min counting), partitioning and rearranging parts. Partitioning involves thinking about numbers as being made up of two parts. Applying this to simple addition, a child may view $2+3$ as $2+2+1$ and solve the problem by skip counting in twos and counting on one. Rearranging parts involves arranging numbers to make computations easier: for example, a child may add $3+5+2$ by adding the $3+2$ first and then recall the answer 10 to solve $5+5$. Around year 2, number fluency encompasses the use of retrieval and decomposition strategies to solve simple addition problems. Around year 3, number fluency involves being able to retrieve all single-digit addition facts and apply these facts to develop efficient mental and written strategies for multi-digit numbers (e.g. $34+42=30+40+4+2$).

While the importance of being able to directly retrieve answers from memory is often emphasised in curriculum documents, in the research literature proficiency is considered to be performance dominated by the accurate use of direct retrieval *and* decomposition strategies (Cowan et al. 2011). The issue associated with achieving proficiency is not whether children learn to use direct retrieval, or direct retrieval and decomposition strategies; the issue is that children no longer rely on counting strategies. Thus, it is important to note that while min counting is the most efficient counting strategy, it is not considered a developmentally appropriate strategy for children to rely on after year 1.

Table 13.1 Problems used at times 1–3, categorised according to problem type

Lower-tie	Upper-tie	Add-to-ten	Near-tie	Add-9	Bridging-10	Small-min
2+2	6+6	2+8	2+3	2+9	3+8	2+4
3+3	7+7	3+7	3+4	3+9	4+7	2+5
4+4	8+8	4+6	4+5	4+9	4+8	2+6
5+5	9+9		5+6	5+9	5+7	2+7
			6+7	6+9	5+8	3+5
			7+8	7+9	6+8	3+6
			8+9			

Note: Shaded problems comprised the 25-problem set used in Study 1. Shaded plus unshaded problems comprised the 36-problem set used in Studies 2 and 3

Capturing Diversity in Simple Addition Performance

We investigated how year 3 children from one primary school solved simple addition problems and conducted three different studies. The school involved is located in an area of mid-socioeconomic status in the Perth metropolitan area. The number of children enrolled in year 3 at the time of the first and second data collection was 61 and 60 children agreed (along with their parents) to participate. The cohort included 30 female and 30 male children with a mean age of 8 years 8 months (standard deviation (*SD*) = 3.6 months) The third data collection involved five children (3 female and 2 male) who were in the school's cohort of year 3 children in the following year. The same procedure for documenting simple addition performance was followed each time and is described below.

Procedure

A researcher (the second author) withdrew each child individually from their class in the morning to a small separate room where they sat together in front of a computer. Initially, the researcher explained to the child that she was interested in how s/he worked out the answer to addition problems and showed the child how the computer program worked. The program randomly displayed a problem from a set of problems, written in the form $x+y=$, where $x \leq y$. The smaller addend was written first to make it possible to distinguish between the use of a min-counting strategy and a less efficient counting-on-from-first strategy. Problems with an addend of one were omitted from the problem set due to difficulties in distinguishing between a retrieval strategy and a count of one. The problem set included 25 problems in study 1 and 36 problems in studies 2 and 3. These are detailed in Table 13.1. A subset was used in study 1 due to time restrictions involved in assessing the whole cohort of year 3 students.

The procedure progressed as follows. The researcher pressed the space bar, which sounded a bell and a timer (unbeknown to the child), and the first problem was presented. The child responded to the problem verbally, where after the researcher immediately pressed the space bar for a second time, stopping the timer and removing the problem from the screen. The child was asked to type his/her answer in the computer. The researcher then asked the child: 'How did you do it?' After describing the strategy they had used to the researcher and following discussion about what the researcher had observed, the child selected the corresponding strategy option on the screen. The options included counting-all, counting-on-from-first, min counting, 'just knew it' (for a retrieval strategy) and 'don't know' (for when the child was unable to explain his/her thinking). When these steps were completed, the process started again for the next problem. Children first practised the procedure with five problems (not in the problem set) before data were collected.

The combination of self-report and observation to identify strategy use on a trial-by-trial basis was recommended by Siegler (1987) and is commonly used to study the strategies used to perform simple addition (e.g. Canobi et al. 1998; Geary et al. 2000). In the studies detailed in this chapter, reaction times (RTs) were used to verify the children's self-reports. Mean RTs to direct retrieval trials in study 1, 2 and 3, respectively, were 2.24 s ($SD=1.18$), 2.48 s ($SD=1.38$) and 2.74 s ($SD=1.19$). Mean RTs to min-counting trials in study 1, 2 and 3, respectively, were 4.87 s ($SD=2.49$), 5.37 s ($SD=2.48$) and 7.51 s ($SD=4.32$). The mean RT to counting-all trials in study 1 (the only study where this strategy was recorded) was 11.97 s ($SD=7.71$). The large *SD* associated with counting trials is easily explained as RTs depend on the number of counts made. The accuracy of children's self-reports were further supported by the clear linear relationship between RTs to min-counting trials where trials were separated according to the minimum addend (i.e. the number of counts made) (see Appendix).

Results: Study 1

The first data collection involved all 60 children solving the problem set on one occasion. In Fig. 13.1, box-and-whisker plots summarise the frequency with which a particular strategy was used by this group of children (both correct and incorrect trials are included). The *stars* and *circles* in Fig. 13.1 are outliers: stars indicate that the frequency for a particular strategy was two times higher than the interquartile range and circles denote that a strategy's frequency was more than 1.5 times higher. The numbers next to these symbols are used in place of children's names.

The boxplot summarising how frequently a count-all strategy was used indicates that use of this strategy was uncommon for the school cohort. Ten children reported some use of the counting-all strategy—six children used it to perform more than 80 % of the problem set (one child, used it exclusively to perform all the problem set) and four children used it at least once but on less than 20 % of the problem set. The boxplot summarising how frequently a counting-on-from-first strategy was used indicates that any use of this strategy was uncommon for the school cohort. Four children used it at least once but on less than 10 % of problems. Two children (specified by the numbers 6 and 47) used both counting-on-from-first and counting-all strategies.

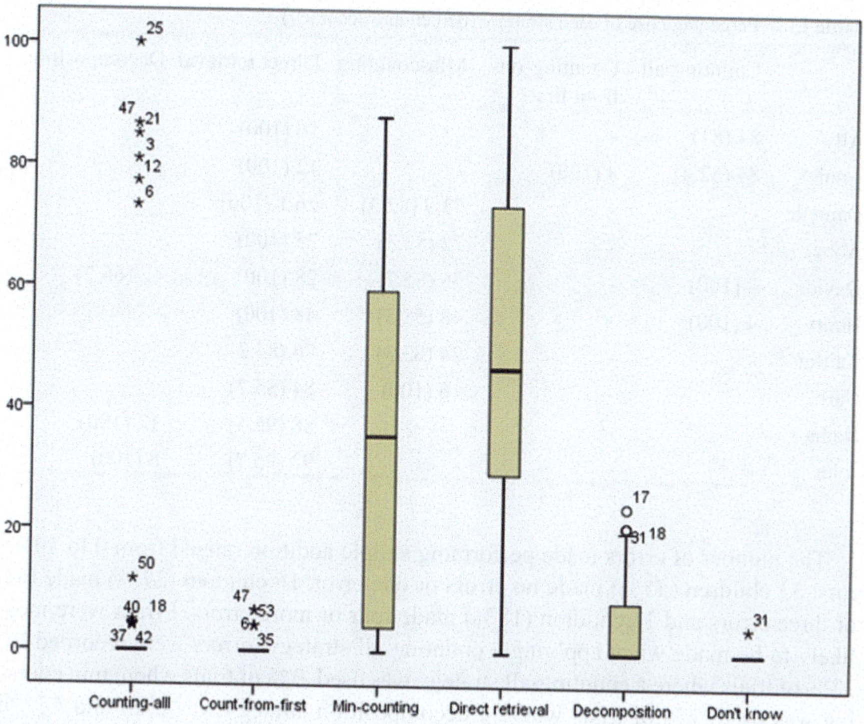

Fig. 13.1 Boxplots depicting the percentage of problems solved using each strategy by the school cohort. Both correct and incorrect trials are included

Boxplots summarising the use of the min-counting, direct retrieval and decomposition strategies indicated that year 3 children at the school generally used a combination of these strategies to perform simple addition: based on the median indicated for each boxplot, half the year 3 cohort performed around 37 % of problems using the min-counting strategy and half performed 46 % of problems using direct retrieval. The range of use of these two strategies among the school cohort was extensive. In regard to the min-counting strategy, the whiskers for the corresponding boxplot indicate that this strategy was used to solve between 0 and 90 % of the problem set; in regard to direct retrieval, the whiskers indicate that this strategy was used to solve between 0 and 100 % of the problem set. Decomposition strategies were generally used less often to solve problems: 0–20 % of the problem set were solved using these strategies.

While the boxplots in Fig. 13.1 indicate that retrieval was the most commonly used strategy among this group of children to perform simple addition, they also reveal that many children were still counting to perform simple addition: 24 children (40 %) were still predominately using min counting to perform simple addition and 6 children (10 %) were predominately using a counting-all strategy.

Table 13.2 Percentage use of each strategy (percentage accuracy)

	Counting-all	Counting-on-from first	Min-counting	Direct retrieval	Decomposition
Ally	84 (81)	-	-	16 (100)	-
Annie	84 (52.4)	4 (100)	-	12 (100)	-
Danielle	-	-	73.9 (83.3)	26.1 (100)	-
Abbey	-	-	72 (83.3)	25 (100)	-
David	4 (100)	-	56 (85.7)	28 (100)	12 (66.7)
Henry	4 (100)	-	48 (58.3)	48 (100)	-
Caitlin	-	-	24 (83.3)	76 (84.2)	
Carl	-	-	16 (100)	84 (85.7)	
Nadia	-	-		88 (95.5)	12 (100)
John	-	-	-	92 (95.7)	8 (100)

The number of errors made performing simple addition ranged from 0 to 10 errors: 33 children (55 %) made no errors or one error, 16 children (27 %) made two or three errors and 11 children (18 %) made four or more errors. Errors were most likely to be made when applying a counting-all strategy: errors were recorded for 27 % of trials where a counting-all strategy was used, 9 % of trials where min counting was used, 7% of trials where a decomposition strategy was used and 4 % of trials where retrieval was used.

To provide a clearer picture of individual variations in children's strategy mix and errors, a purposeful sampling procedure was used to choose 8 out of the 55 children who showed a more typical pattern of strategy use for the school cohort. Two children from each quartile represented on the min-counting boxplot were randomly selected. Two children were also randomly selected from the group of six children who relied heavily on a counting-all strategy. The strategy mix for the 10 children selected along with the accuracy with which each strategy was performed is shown in Table 13.2.

The figures in Table 13.2 highlight the variations in the strategy mix for each child and capture some of the diversity their teachers contend with each day in the classroom. For example, Annie predominantly solved simple addition problems by counting from one and was highly inaccurate (performing only half of these problems correctly). In comparison, Nadia and John performed all the problems using a combination of direct retrieval and decomposition strategies (and were mostly accurate). Different patterns of inaccuracy also emerged: Henry, for example, was often inaccurate when applying the min-counting strategy but was always accurate when using directly retrieval; Carl was always accurate using the min-counting strategy but was sometimes inaccurate using direct retrieval.

The Australian curriculum document suggests that by around year 3, retrieval does not just dominate performance but it is used exclusively to perform simple addition. The findings from study 1 indicate that from this school, only 20 % of the year 3 cohort was close to meeting this expectation. Around 40% of cohort solved the majority of simple addition problems using min counting and 10% used a counting-all strategy.

It is important to acknowledge a limitation of this study—children's performance was assessed at only one point in time. We felt it would be too easy to dismiss the findings from study 1 because of this limitation and so conducted a second study. In study 2, we selected a sample of nine children and tracked their performance on 15 occasions. Findings for three of these children are presented in the next section.

Results: Study 2

Assessing a child at one time gives only a limited picture of how they perform simple addition. It is well documented that children, who are still developing proficiency with certain skills, including simple addition, will often use different strategies to solve the same problem on different occasions (Siegler 1995). This means that children will sometimes use an inefficient strategy even though on a previous occasion they have shown they can solve the same problem using a more efficient strategy. Inefficient strategies will compete with efficient strategies for selection until gradually efficient strategies dominate performance (Siegler 1995). There are other reasons why assessing a child at one time provides only a limited picture. One assessment is unable to suggest if inaccurate performance is due to the novelty of the assessment task or if errors are repeated or are inconsistent over time. Consistent errors would be suggestive of a procedural 'bug' where a strategy is misapplied because of a misconception or faulty rule. Nine children who participated in study 1 were purposefully selected to participate in a second study where their performance was assessed each day for 15 days. The results of three children whose names are highlighted in Table 13.2 are reported in this chapter to illustrate how children's strategy mix varied over time. It is important to note that in study 1, these three children predominantly used the min-counting strategy to solve simple addition problems. Self-reports of strategy use in study 1 were consistent with self-reports of strategy use in study 2.

Area graphs depicted in Fig. 13.2 illustrate how the three children's strategy mix varied over the 15 days. Only correct trials are included in these graphs. A horizontal line drawn to represent 36 problems in Fig. 13.2 is used to indicate the number of errors made at each time interval. The distance between the top of the area graph and the line indicates the number of incorrect trials. Given that performance on 15 occasions was documented and that each time the assessment task was completed a child practised their skills with simple addition, these graphs also depict how the strategy mix changed as a result of *extended practice*. The term extended practice denotes practice that is concentrated and frequent. During this time of practice, no child received feedback on the accuracy of their performance.

The area graphs in Fig. 13.2 show marked patterns of individual variability in strategy use on different occasions. While this is not a new finding, these graphs

<table>
<tr><th>Name</th><th colspan="2">Known key facts (in black)</th><th>Variations in strategy mix over time</th></tr>
<tr><td>Danielle</td><td>2+2
2+8
3+3
3+7
4+4
4+6</td><td>5+5
6+6
7+7
8+8
9+9</td><td>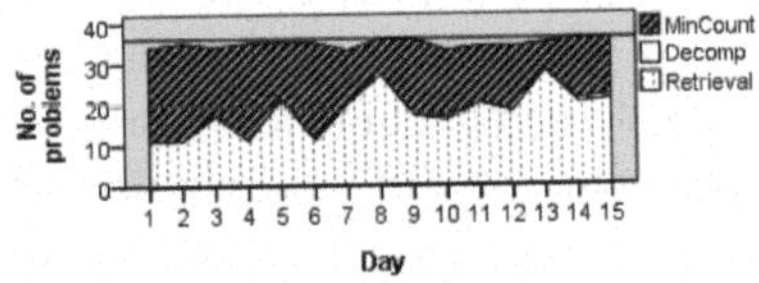
</td></tr>
<tr><td>Abbey</td><td>2+2
2+8
3+3
3+7
4+4
4+6</td><td>5+5
6+6
7+7
8+8
9+9</td><td>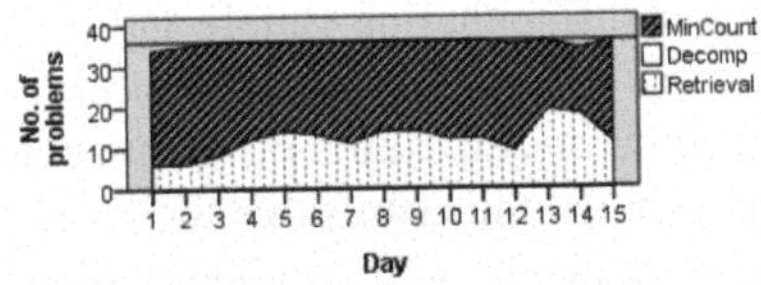
</td></tr>
<tr><td>David</td><td>2+2
2+8
3+3
3+7
4+4
4+6</td><td>5+5
6+6
7+7
8+8
9+9</td><td>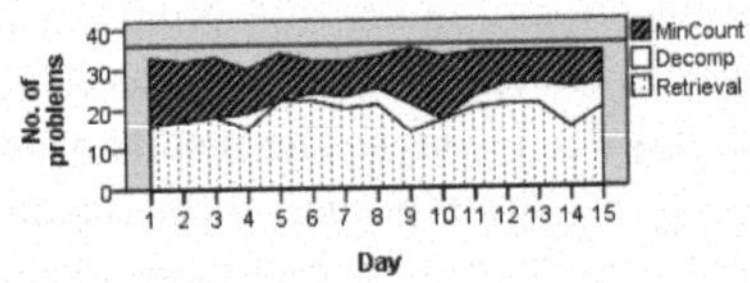
</td></tr>
</table>

Fig. 13.2 Area graphs depicting changes in children's strategy mix over time in study 2

are a novel way of illustrating this variability in strategy use over time and with practice. These graphs also highlight the important role practice has on developing proficiency with simple addition. As a result of extended practice, Danielle, Abbey and David use min counting less frequently and retrieval more frequently to perform simple addition. David also increased his use of decomposition strategies.

As strategy use on different occasions was documented, it was also possible to investigate what key *facts* were well known by each child. A fact is considered a key fact if it is a tie fact or an add-to-ten fact, as these facts are used most com-

Table 13.3 The number of repeated (R) and non-repeated (NR) errors by each participant in study 2

	Min counting (min)			Retrieval (ret)			Decomposition (decomp)		
	NR	R	Total (% of min trials)	NR	R	Total (% of ret trials)	NR	R	Total (% of decomp trials
Danielle	7	3	10 (3.9%)	4	4	8 (2.9%)	–	–	–
Abbey	3	2	5 (1.4%)	1	–	1 (0.6%)	–	–	–
David	10	4	14 (7.6%)	12	5	17 (5.7%)	8	2	10 (16.9%)

monly in decomposition strategies. A child was considered to have known a fact if they directly retrieved the correct answer on each occasion over the first five days. Each child's known key facts are indicated in Fig. 13.2. Children who initially used direct retrieval less often had not established all the tie facts (particularly the higher tie facts) and many of the add-to-ten facts. Given that initially these children were not retrieving these key facts, it is not surprising that their use of decomposition strategies was limited.

As errors on different occasions were documented, the number of repeated and non-repeated errors could also be explored. Errors were coded as repeated errors if a child produced the same incorrect answer for a problem using the same strategy on more than one occasion. Each child's errors are categorised in Table 13.3 according to strategy use.

Each child's repeated errors were scrutinised to find patterns that suggested a procedural bug where the min-counting strategy was misapplied due to a misconception. No procedural bugs were detected. It is noteworthy that Abbey produced the least number of errors and displayed the lowest use of direct retrieval. Abbey's performance reflects what has been referred to as *perfectionist* performance (Siegler 1988), which is highly accurate performance that is dominated by counting. This pattern of performance is discussed in more detail in the final section of this chapter.

The strategy choice model predicts that some children will continue to use a counting strategy for a longer period of time because they often make mistakes during the counting procedure. As inaccuracy was not a criterion for selecting the nine children in study 2, we did not have a clear picture of the effects of practice on children's performance if they were prone to making errors. (David made the most number of errors of the nine participants, in total 41 errors over 15 days. This represents an average of less than three errors on the problem set. The results of study 1 indicated that some children were making up to 10 errors). We decided to conduct a third study to investigate if extended practice benefitted children who were prone to making errors. To do this, we purposefully selected five children from the year 3 cohort in the following year. These children were identified by their teacher as being reliant on min counting and as often making mistakes with simple addition.

Results: Study 3

Five children were tracked every day for eight consecutive school days and each day they performed the set of 36 simple addition problems. Due to time restrictions, each child's performance was tracked for a shorter period than in study 2. Variations in each child's strategy mix over the 8 days are illustrated in Fig. 13.3 along with key facts that each child was consistently and correctly retrieving.

Although the shorter tracking period made comparisons between children in study 2 and study 3 problematic, the results suggest that extended practice was not as effective at increasing the use of direct retrieval for children in study 3 as it was for children in study 2. This was deduced by drawing a line of best fit through the number of problems correctly retrieved over the period of extended practice for each child. The slopes of the lines for children in study 3 were steeper than those for children in study 2. Most notably, practice had little effect on increasing the frequency of retrieval for Bethany, Jane and Mike.

The numbers of errors recorded for children in study 3 were generally higher than those recorded for children in study 2. This is noticeable by comparing the gap from each area graph and the horizontal line drawn at 36 (denoting 36 problems). A summary of the number of non-repeated and repeated errors are given in Table 13.4.

As highlighted in Table 13.4, there were more non-repeated min-counting errors than repeated min-counting errors. An examination of the non-repeated counting errors revealed that these children often overcounted by one, undercounted by one or started at and counted on the same addend (for example Mike calculated that $5+9=18$ and Bethany calculated that $4+8$ was 8). These errors suggested that children were losing track of the count during the counting procedure. Sometimes they would overcount, sometimes they would undercount; other times they would end up counting on the wrong addend. Non-repeated retrieval errors were also often over by one, under by one or double one of the addends (matching min-counting errors).

Repeated min-counting errors were scrutinised for evidence of a procedural bug. A pattern was found in Jane's performance where she often confused a 6 with a 9. Repeated errors that were directly retrieved were also scrutinised. An obvious pattern emerged for Mike: he applied a faulty rule to tie facts stating that $4+4=14$, $6+6=16$, $7+7=17$, $8+8=18$, $9+9=19$. Also, both Bethany and Harry consistently retrieved the answer 14 for the problem $8+8$.

The findings from study 3 suggest that practice is not as effective at increasing the likelihood of retrieval for children who are prone to making errors. A more detailed analysis of study 3 data is presented in de Villiers and Hopkins (2013).

Discussion

Curriculum documents suggest that by year 3, children are able to accurately recall simple addition facts. Findings from time 1 indicated that only 20% of year 3 children in the school met or were close to meeting this expectation. This finding is not

<table>
<tr><th>Name</th><th>Known key facts (in black)</th><th>Variations in strategy mix over time</th></tr>
<tr><td>Ainsley</td><td>2+2 5+5
2+8 6+6
3+3 7+7
3+7 8+8
4+4 9+9
4+6</td><td>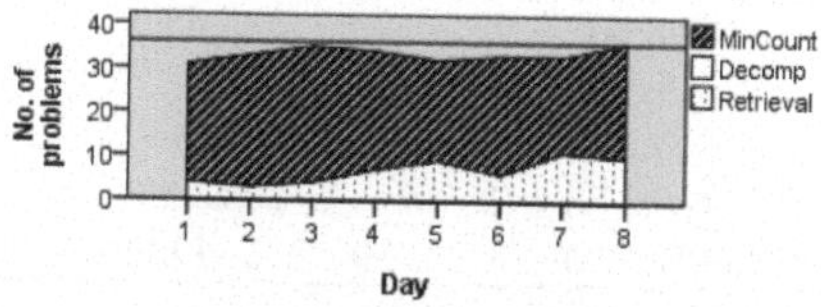
</td></tr>
<tr><td>Bethany</td><td>2+2 5+5
2+8 6+6
3+3 7+7
3+7 8+8
4+4 9+9
4+6</td><td>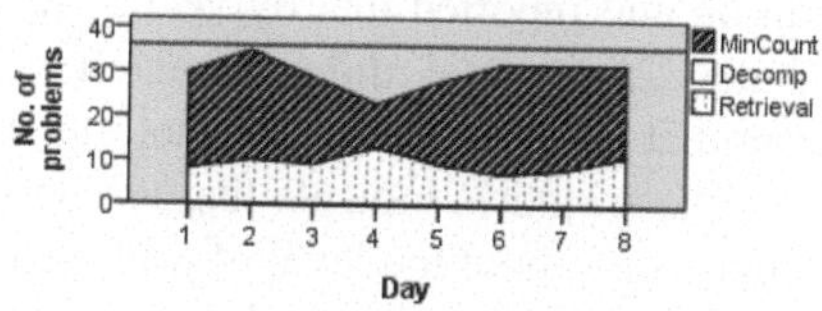
</td></tr>
<tr><td>Harry</td><td>2+2 5+5
2+8 6+6
3+3 7+7
3+7 8+8
4+4 9+9
4+6</td><td>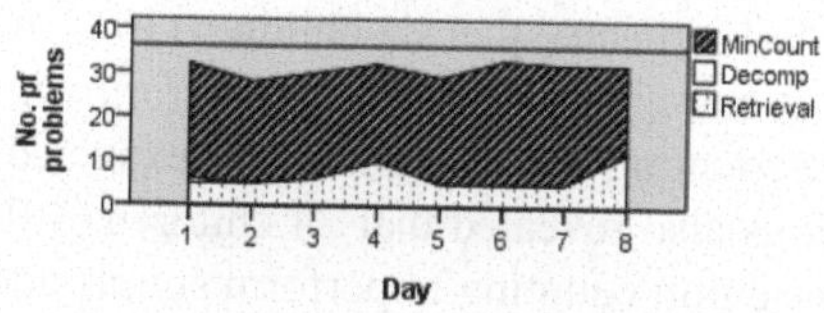
</td></tr>
<tr><td>Jane</td><td>2+2 5+5
2+8 6+6
3+3 7+7
3+7 8+8
4+4 9+9
4+6</td><td>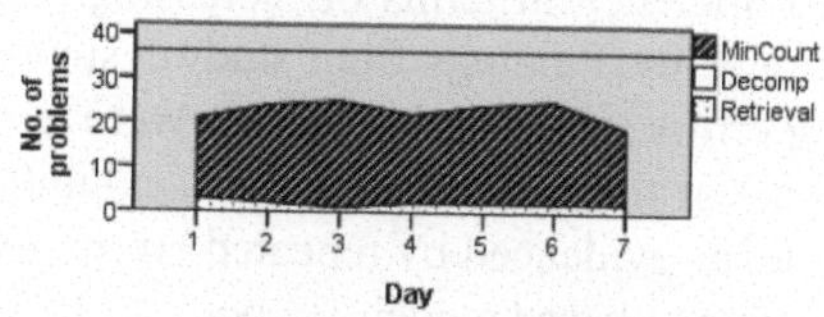
</td></tr>
<tr><td>Mike</td><td>2+2 5+5
2+8 6+6
3+3 7+7
3+7 8+8
4+4 9+9
4+6</td><td>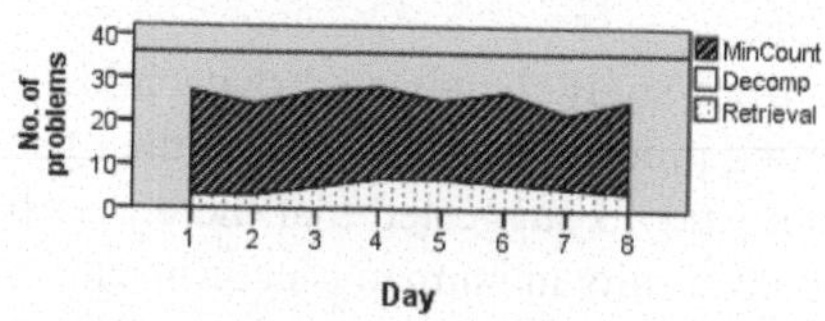
</td></tr>
</table>

Fig. 13.3 Area graphs depicting changes in children's strategy mix over time in study 3. Jane was absent from school and was not tracked on the eighth day

Table 13.4 The number of non-repeated (NR) and repeated (R) errors by each participant in study 3

	Min counting (min)			Direct retrieval (ret)			Decomposition (decomp)		
	NR	R	Total (% of min trials)	NR	R	Total (% of ret trials)	NR	R	Total (% of decomp trials)
Ainsley	10	8	18 (7.8 %)	2	–	2 (3.6 %)	–	–	–
Bethany	18	12	30 (15.3 %)	7	3	10 (11.8 %)	–	–	–
Harry	10	12	22 (10.2 %)	1	13	14 (20.9 %)	2	0	2 (66.7 %)
Jane	50	32	82 (36.0 %)	2	–	2 (12.5 %)	–	–	–
Mike	29	14	43 (21.1 %)	8	30	38 (49.4 %)	–	–	–

unique to the school. Cowan et al. (2011) investigated how children in UK schools performed simple addition and reported that by year 3, few children retrieved the answers to most problems. Cowan et al. did not make it clear if educators need be concerned about the fact that many children were not meeting curriculum expectations. The findings from these studies indicate that teachers from the participating school *do* need to be concerned about the level of proficiency displayed by at least half their year 3 cohort. As there is no hint that this primary school is unique in how children are taught mathematics, we were surprised to find just how many children were not showing signs of proficiency with simple addition.

Findings from study 1 indicated that six children (10 % of the year 3 cohort) were predominately using a counting-all strategy to perform simple addition. This is a particularly immature counting strategy, one that is expected to have disappeared by year 1. Study 1 findings also revealed that 24 children (40 % of the year 3 cohort) still predominately used min counting to perform simple addition.

Study 2 findings indicated that for some children, a lack of proficiency could be addressed with extended practice. Extended practice was effective in improving some children's proficiency in terms of increasing the use of direct retrieval and decomposition strategies. Findings from study 3 suggested that children who were prone to making errors with simple addition were less likely to benefit from extended practice. Some of the errors that children exhibited were due to a procedural bug or faulty rule, as evidenced by repeated errors. Evidence of a faulty rule is consistent with the schema-based model (Baroody 1994; Baroody and Tiilikainen 2003) of development given that a faulty rule would have been constructed by the learner. Non-repeated errors were also common, suggesting that participants were losing track of the count.

Whereas findings from study 2 indicated that children who occasionally lost track during the counting procedure still increased their use of direct retrieval with practice, findings from study 3 suggested that there may be a tipping point when losing track of the count results in so many inconsistent answers that practice has little effect on promoting the use of direct retrieval. These findings are consistent with predictions based on the strategy choice model (Shrager and Siegler 1998; Siegler and Shipley 1995; Siegler and Shrager 1984). If practice with a counting backup strategy results in the same error being made on the same problem, then the association between a problem and the wrong answer will strengthen in memory, increasing the likelihood that this wrong answer is retrieved in the future. However,

Table 13.5 Patterns of difficulty

Pattern	Description of simple addition performance
Inefficient counting	Dominated by the use of inefficient counting strategies—predominantly a counting-all strategy. Inaccurate performance is common
Effective counting	Dominated by the use of a min-counting strategy but generally accurate performance—few correct key facts are consistently retrieved
Losing track of the count	Dominated by the use of min counting with non-repeated min-counting errors. The extent of these errors appears to be a tipping point affecting the benefits of practice to promote retrieval
Guessing	A high number of non-repeated direct retrieval errors
Procedural bug	Some use of min counting with repeated min-counting errors
Faulty rule	A high number of repeated direct retrieval errors

if practice results in different wrong answers being associated with the same problem, then retrieval will not occur but a counting backup strategy will continue to be relied upon. Children in study 3 frequently made min-counting errors that were not repeated and they continued to rely on min counting.

The strategy choice model also predicts that some children will continue to use a counting strategy for a longer period of time than their peers because they have a high confidence criterion or threshold for retrieving answers (Shrager and Siegler 1988). In other words, they need more practice using a counting strategy before they are likely to trust that they just know the answer, even though they are mostly accurate each time they use a counting strategy. This pattern of performance, referred to as effective counting in Table 13.5, was illustrated by Abbey in study 2 (and is referred to as perfectionist performance elsewhere). In summary, six patterns of difficulty were noted for this cohort of children and are summarised in Table 13.5.

Published interventions that have focused on improving children's simple addition performance have largely relied on three approaches: extended practice (e.g. Lin et al. 1994), direct instruction of efficient strategies including min counting (Tournaki 2003; Fuschs et al. 2010) *or* fact memorisation (Fuschs et al. 2006; Poncy et al. 2007). Based on the findings presented here, it is possible to reason which children are more likely to benefit from extended practice or direct instruction, and which children may actually be disadvantaged by a particular approach.

Extended practice is unlikely to benefit children who display inefficient counting given they are already more than 2 years behind their peers. These children are more likely to benefit from being explicitly taught the min-counting strategy along with the concepts and skills that underpin use of this strategy. Extended practice is likely to benefit children who show a pattern of effective counting or who sometimes lose track of the count. However, if children are often losing track of the count, extended practice is likely to be less effective. It seems critical that practice for children who often lose track of the count is closely monitored. We recommend that these children receive feedback on their accuracy immediately after each problem and are given the opportunity to correct their mistake. Research is needed to ascertain if practice that is monitored in this way improves the likelihood of retrieval. Children who have developed a procedural bug or faulty rule are likely to be disadvantaged by a program of extended practice as this could result in the strengthening of associations in memory between problems and incorrect answers—resulting in retrieval errors. We recommend that an intervention must first address these misconceptions before extended practice is required.

We recommend that children who display effective counting are also explicitly taught decomposition strategies, as they have a high confidence criterion for trusting retrieval. Given that decomposition strategies make use of certain retrieved facts, it will be important to first ascertain what facts children do retrieve so that they can be taught decomposition strategies that make use of these known facts.

Different patterns of difficulty suggest that different approaches are needed to help children develop proficiency with simple addition. While further research is needed to explore the generalizability of these patterns of difficulty presented here, the approach taken to capture the diversity in how children perform simple addition provides much helpful information to the teachers in the participating school. We recommend that all lessons have a section where children have an opportunity to improve their number fluency. As children develop proficiency with simple addition and subtraction then developing efficient mental and written strategies for multi-digit addition and subtraction, would be appropriate skills to work on. We do not recommend that children are streamed or removed from class to develop an aspect of number fluency, nor do we recommend that number fluency be the only focus of a particular lesson—(for reasons outlined by Peter Sullivan in Chapter 14). We find that many teachers introduce their lessons with 'mental mathematics' and recommend that these activities are targeted to meet individual children's fluency needs.

Appendix

Reaction times to min-counting trials separated by the number of counts required

Minimum addend	Study 1		Study 2		Study 3	
	Mean (s) (SD)	*N*	Mean (s) (SD)	*N*	Mean (s) (SD)	*N*
2	3.55 (2.11)	86	3.85 (1.96)	103	4.62 (2.18)	190
3	3.89 (1.49)	146	4.28 (1.68)	206	5.59 (2.57)	211
4	4.77 (2.12)	119	4.94 (1.96)	254	7.26 (3.33)	200
5	5.78 (2.27)	70	5.48 (2.13)	190	7.99 (3.71)	156
6	6.30 (3.74)	41	6.15 (2.58)	180	10.05 (5.42)	131
7	6.53 (2.22)	46	6.83 (2.48)	125	10.51 (4.89)	105
8	6.38 (2.35)	17	7.57 (3.75)	65	10.29 (4.59)	63
9	8.62 (2.84)	12	6.06 (2.87)	8	11.72 (5.72)	24

Note: The size of the SD associated with mean RTs is likely to be influenced by trials where the counting procedure is interrupted and/or counts are repeated due to self-correction, as documented in Hopkins and Lawson (2002). This argument is supported by the general pattern that SDs increased as the minimum addend increased (children are more likely to lose track when making more counts) and that the highest SDs were recorded in study 3 where children were more likely to lose track and make errors during the min-counting procedure

SD standard deviation

References

Ashcraft, M. H. (1995). Cognitive psychology and simple arithmetic: A review and summary of new directions. *Mathematical Cognition, 1*(1), 3–34.

Australian Curriculum, Assessment and Reporting Authority (ACARA). (2010). Australian curriculum V1.2—mathematics: Content structure. http://www.australiancurriculum.edu.au/Mathematics/Content-structure. Retrieved from 25 June 2014.

Baroody, A. J. (1994). An evaluation of evidence supporting fact-retrieval models. *Learning and Individual Differences, 6*(10), 1–36.

Baroody, A. J., & Dowker, A. (Eds.). (2003). *The development of arithmetic concepts and skills: Constructing adaptive expertise*. Mahwah: Lawrence Erlbaum.

Baroody, A. J., & Tiilikainen, S. P. (2003). Two perspectives on addition development. In A. Baroody & A. Dowker (Eds.), *The development of arithmetic concepts and skills: Constructing adaptive expertise* (pp. 75–125). Mahwah: Lawrence Erlbaum.

Bull, R., & Johnston, R. S. (1997). Children's arithmetical difficulties: Contributions from processing speed, item identification, and short-term memory. *Journal of Experimental Child Psychology, 65,* 1–24.

Canobi, K. H., (2009). Concept-procedure interactions in children's addition and subtraction. *Journal of Experimental Child Psychology, 102,* 131–149.

Canobi, K. H., Reeve, R. A., & Pattison, P. E. (1998). The role of conceptual understanding in children's addition problem solving. *Developmental Psychology, 34*(5), 882–891.

Carpenter, T. P., & Moser, J. M. (1984). The acquisition of addition and subtraction concepts in grades one through to three. *Journal for Research in Mathematics Education, 15,* 179–202.

Christensen, C. A., & Cooper, T. J. (1991). The effectiveness of instruction in cognitive strategies in developing proficiency in single-digit addition. *Cognition and Instruction, 8*(4), 363–371.

Cowan, R., Donlan, C., Shepherd, D., Cole-Fletcher, R., Saxton, M., & Hurry, J. (2011). Basic calculation proficiency and mathematics achievement in elementary school children. *Journal of Educational Psychology, 103*(4), 786–803.

Cummings, J. J. (1988). *Children's cognitive performance in the addition facts and more complex sums: The development and importance of automaticity*. Unpublished thesis.

Department for Education. (2013). The UK national curriculum framework. http://www.gov.uk/dfe. Retrieved from 25 June 2014.

de Villiers, C., & Hopkins, S. (2013). When practice doesn't lead to retrieval: An analysis of children's errors with simple addition. In V. Steinle, L. Ball, & C. Bardini (Eds.), *Mathematics education: Yesterday, today and tomorrow* (Proceedings of the 36th annual conference of the Mathematics Education Research Group of Australasia). Melbourne: MERGA.

Fuchs, L. S., Fuchs, D., Hamlet, C. L., Powell, S. R., Capizzi, A. M., & Seethaler, P. M. (2006). The effects of computer-assisted instruction on number combination skill in at-risk first graders. *Journal of Learning Disabilities, 39*(5), 467–475.

Fuchs, L. S., Powell, S. R., Seethaler, P. M., Cirino, P. T., Fletcher, J. M., Fuchs, D., & Hamlett, C. L. (2010). The effects of strategic counting instruction, with and without deliberate practice, on number combination skill among students with mathematics difficulties. *Learning and Individual Differences, 20*(2), 89–100.

Geary, D. C. (2010). Missouri longitudinal study of mathematical development and disability. In R. Cowan, M. Saxton, & A. Tolmie (Eds.), *Understanding number development and number difficulties (No. 7, British Journal of Educational Psychology, Monograph Series II: Psychological Aspects of Education—Current Trends*, pp. 31–49). Leicester: British Psychological Society.

Geary, D. C., Hamson, C. O., & Hoard, M. K. (2000). Numerical and arithmetical cognition: A longitudinal study of process and concept deficits in learning disabled children. *Journal of Experimental Child Psychology, 77*(3), 236–263.

Gilmore, C. K., & Papadatou-Pastou, M. (2009). Patterns of individual differences in conceptual understanding and arithmetical skill: A meta-analysis. *Mathematical Thinking and Learning, 11,* 25–40.
Hopkins, S. L., & Lawson, M. J. (2002). Explaining the acquisition of a complex skill: Methodological and theoretical considerations uncovered in the study of simple addition and the moving-on process. *Educational Psychology Review, 14,* 121–154.
Jordan, N. C., & Oettinger-Montani, T. (1997). Cognitive arithmetic and problem solving: A comparison of children with specific and general mathematics difficulties. *Journal of Learning Disabilities, 30*(6), 624–634.
Jordan, N. C., Hanich, L. B., & Kaplan, D. (2003). A longitudinal study of mathematical competencies in children with specific mathematics difficulties versus children with comorbid mathematics and reading difficulties. *Child Development, 74*(3), 834–850.
Landerl, K., Bevan, A., & Butterworth, B. (2004). Developmental dyscalculia and basic numerical capacities: A study of 8–9-year-old students. *Cognition, 93,* 99–125.
Lin, A., Podell, D. M., & Tournaki-Rein, N. (1994). CAI and the development of automaticity in mathematics skills in students with and without mild mental handicaps. *Computers in the Schools, 11,* 43–58.
McIntosh, A., Reys, B. J., & Reys, R. E. (1992). A proposed framework for examining basic number sense. *For the Learning of Mathematics, 12*(3), 2–8.
National Council of Teachers of Mathematics. (2000). *Principles and standards for school mathematics.* Reston: NCTM.
Ostad, S. A. (1997). Developmental differences in addition strategies: A comparison of mathematically disabled and mathematically normal children. *British Journal of Educational Psychology, 67*(3), 345–357.
Poncy, B. C., Skinner, C. H., & Jaspers, K. J. (2007). Evaluating and comparing interventions designed to enhance math fact accuracy and fluency: Cover, copy, and compare versus taped problems. *Journal of Behavioral Education, 16,* 27–37.
Rathmell, E. C. (1978). Using thinking strategies to teach the basic facts. In M. N. Suydam & R. E. Reys (Eds.), *Developing computational skills: 1978 yearbook.* Reston: NCTM.
Robinson, C. S., Menchetti, B. M., & Torgesen, J. K. (2002). Toward a two-factor theory of one type of mathematics disabilities. *Learning Disabilities Research & Practice, 17,* 81–89.
Schneider, M., & Stern, E. (2010). The developmental relations between conceptual and procedural knowledge: A multimethod approach. *Developmental Psychology, 46*(1), 178–192.
Shrager, J., & Siegler, R. S. (1998). SCADS: A model of children's strategy choices and strategy discoveries. *Psychological Science, 9*(5), 405–410.
Siegler, R. S. (1987). The perils of averaging data over strategies: An example from children's addition. *Journal of Experimental Psychology: General, 116*(3), 250–264.
Siegler, R. S. (1988). Individual differences in strategy choices: Good students, no-so-good students and perfectionists. *Child Development, 59*(4), 833–851.
Siegler, R. S. (1995). How does change occur: A microgenetic study of number conservation. *Cognitive Psychology, 28*(3), 225–273.
Siegler, R. S., & Jenkins, E. (1989). *How children discover new strategies.* Hillsdale: Erlbaum.
Siegler, R. S., & Shipley, C. (1995). Variation, selection and cognitive change. In T. Simon & G. S. Halford (Eds.), *Developing cognitive competence: New approaches to process modelling.* Hillsdale: Erlbaum.
Siegler, R. S., & Shrager, J. (1984). Strategy choices in addition and subtraction: How do children know what to do? In C. Sophian (Ed.), *Origins of cognitive skills* (pp. 229–293). Hillsdale: Erlbaum.
Steinberg, R. M. (1985). Instruction on derived fact strategies in addition and subtraction. *Journal for Research in Mathematics Education, 16,* 337–355.
Swanson, H. L., & Jerman, O. (2006). Math disabilities: A selective met-analysis of the literature. *Review of Educational Research, 76*(2), 249–274.

Torbeyns, J., Verschaffel, L., & Ghesquière, P. (2004). Strategy development in children with mathematical disabilities: Insights from the choice/no-choice method and the chronological-age/ability/level-match design. *Journal of Learning Disabilities, 37,* 119–131.

Tournaki, T. (2003). The differential effect of teaching addition through strategy instruction versus drill and practice to students with and without learning disabilities. *Journal of Learning Disabilities, 36,* 449–458.

Chapter 14
Maximising Opportunities in Mathematics for All Students: Addressing Within-School and Within-Class Differences

Peter Sullivan

Introduction

This chapter argues that the highest quality teaching maximises the learning of all students, not just a lucky few, and that teachers, schools and systems should address structural and other factors that might inhibit the goal of creating opportunities for all students. It begins with an overview of the aspirations of the Australian Curriculum Assessment and Reporting Agency (ACARA) for curriculum and schooling in Australia, then describes some of the challenges facing Australian schools, continues by arguing that the major issue is within-school differences, describes school and classroom grouping practices and concludes with an outline of an approach that seems to make heterogeneous grouping feasible. In the discussion, the focus is mainly on the mathematics curriculum and teaching, in part because it is the subject that seems to experience the greatest challenge due to differences in student readiness, and in part because closing off options for mathematics study restricts both study and employment options for students. While the focus of the discussion is on Australian contexts and Australian schools, the implications and conclusions are relevant internationally. In a sense, the focus on Australia can be taken as a case study.

Aspirations for Education in Australia and the Australian Curriculum

Fundamental to system, school and classroom decisions on maximising opportunity are the potential of education to create opportunity for citizens that they might not otherwise have. The commitment of government through its various agencies is

---

P. Sullivan (✉)
Faculty of Education, Monash University, Clayton, Australia
e-mail: peter.sullivan@monash.edu

A. Bishop et al. (eds.), *Diversity in Mathematics Education*,
Mathematics Education Library, DOI 10.1007/978-3-319-05978-5_14,

unambiguous, as evident in the overarching Shape Paper (ACARA 2012) that established the principles for the Australian Curriculum (AC):

> All Australian governments have committed to the goals of the Melbourne Declaration, which are that Australian schooling promotes equity and excellence; and that all young Australians become successful learners, confident and creative individuals, and active and informed citizens. (p. 5)

The Shape Paper went further to argue that schooling and the curriculum should ensure that young people

> have a sense of self-worth, self-awareness and personal identity that enables them to manage their emotional, mental, spiritual and physical wellbeing. (p. 8)

This, in turn, is intended to prepare them for their

> potential life roles as family, community and workforce members, [so they will be able to] embrace opportunities, make rational and informed decisions about their own lives and accept responsibility for their own actions. (p. 9)

This is even described as an *entitlement*

> of each student to knowledge, skills and understandings that provide a foundation for successful and lifelong learning and participation in the Australian community. (p. 10)

The document also makes the explicit assumption

> that each student can learn and the needs of every student are important. It enables high expectations to be set for each student as teachers account for the current levels of learning of individual students and the different rates at which students develop. (p. 10)

Similar sentiments are expressed in the principles for mathematics (ACARA 2009):

> Building on the draft National Declaration on Educational Goals for Young Australians, a fundamental aim of the mathematics curriculum is to educate students to be active, thinking citizens, interpreting the world mathematically, and using mathematics to help form their predictions and decisions about personal and financial priorities. Mathematics also enables and enriches study and practice in many other disciplines. (p. 5)

It also argues

> that schooling should create opportunities for every student. There are two aspects to this. One is the need to ensure that options for every student are preserved as long as possible, given the obvious critical importance of mathematics achievement in providing access to further study and employment and in developing numerate citizens. (p. 10)

In other words, the documents which can be taken to represent community aspirations, argue that all students have an entitlement to a curriculum that maximises their opportunities, that prepares them for a life in which creativity, imagination and an orientation to life-long learning are emphasised more than correct answers, compliant attitudes and acceptance of a designated place in a hierarchical social order. The assumption is that schools and classrooms should be structured to facilitate the achievement of the former curriculum goals for all students.

Challenges Facing Australian Schools in Maximising Opportunities for All

There are a number of factors that make the full implementation of a curriculum based on these principles difficult and urgent. One factor is the long 'tail' (McGaw 2007) that refers to the fact that while some Australian students are doing well in international comparisons, there are other students who are a long way behind them in readiness for further study.

The characteristics of this tail are elaborated by Thompson et al. (2010) who note that while the performance of Australian students in mathematics in PISA 2009 had remained strong in comparison to previous surveys, the ranking of the full cohort of Australian students in mathematics had declined, and this decline was reported as mainly due to a fall in the proportion of students achieving at the top levels. They also note that students in the lower groups were disproportionally those from

- the lowest socioeconomic status (SES) quartile (of whom 23 % were not achieving level 2 in literacy compared with 5 % of the high SES background students, with the figures for numeracy being 22 and 5 %, respectively);
- indigenous background (of whom 38 % were not achieving level 2 in literacy compared with 12 % of non-Indigenous students, with the figures for numeracy being 39 and 12 %, respectively); and
- rural areas (of whom 24 % were not achieving level 2 in literacy compared with 12 % of metropolitan students, with the figures for numeracy being 28 and 12 %, respectively).

Importantly, in the case of numeracy, students not achieving level 2 are not yet able to use basic procedures or interpret results. These students would experience substantial difficulty with the mathematics curriculum relevant for their age and year level.

A key challenge for schools is to find ways to address the needs of these students. But a critical consideration is that within each of these subgroups there is widespread diversity. In elaborating this issue, Thompson et al. (2010) compare the reading levels of Australian-born, first generation, and foreign-born students. There was a slight advantage to the first generation students, but the real issue is that the diversity of achievement in each of the three groups of students was more or less identical, in that there were similar proportions of students from each group at each of the achievement levels as defined by PISA. Similarly, there is a diversity of achievement of low SES group students, Indigenous students and rural students, with students in each of these subgroups achieving at the top international levels. The challenge for schools is to ensure that students from these subgroups are not given a restricted curriculum but to find ways to address the diversity of readiness whatever the student population.

In other words, even though there is a long tail in achievement of Australian students, with particular subgroups over-represented in the tail, there are also students even within the subgroups achieving at the best international levels. These factors all make the task of structuring schools and classrooms to maximise opportunity challenging.

Between- and Within-School Differences

Even when acknowledging the disparity between the resourcing of schools in different sectors and the differences in status of government schools depending on their location, it seems that the major variances are within individual schools rather than between schools. In an Australian Council of Educational Research report on students' tertiary entrance scores, Marks et al. (2001) argue that between-school differences account for approximately 22 % of the variation in student scores, and that about half of this variation is accounted for by differences in the academic and socioeconomic mix of students and school sector. In other words, 78 % of the variance in student scores is attributable to differences within each school. Similarly, in an analysis of systemic assessment results, Rothman and McMillan (2003) argue that less than 20 % of the variation in achievement scores on both reading and mathematics could be attributed to differences between schools. As with Marks et al., they note that around half of the between-school variance could be explained by student characteristics. In other words, the challenge facing Australian schools is not differences between school types and suburbs, but the wide range of achievement *within* each school, and that actions to maximise opportunity are the responsibility of individual schools and classroom teachers.

The inference is that differences within individual schools should be the focus of system and school policy development. These within-school differences related predominantly to differences between teachers and grouping practices that either minimise or exacerbate the differences.

Accommodating the Diversity of Readiness in School and Classroom Grouping Structures

It is no simple task to address this diversity of readiness and the challenges teachers face, particularly in years 5–10. Teachers at those levels are more likely than others to experience classrooms in which there are:

- Fast learners who shout answers and criticise others who are still thinking through problems that the fast learners have already solved, and who complain to their parents about being under-extended.
- Some other learners who have more or less given up believing that they cannot learn, and who prefer to interrupt others.
- Extensive and exhaustive lists of content to cover that pressure teachers to move quickly from topic to topic.

- Routines in schools that leave teachers with limited time for collaboration, sharing ideas, innovating, resource development and so on.

In other words, there are very real pressures on teachers that are directly relevant to their approaches to addressing differences in readiness for the curriculum at particular levels. The focus in this section is on the ways that schools respond to these pressures and in particular on the decisions about the ways students are grouped. Such decisions are often made based on the preferences of the teachers, which can be informed by their views on who can learn and also by their concerns for particular categories of students that they feel might be disadvantaged by certain school and class groupings. These issues are especially acute in the case of mathematics as the stratification of groups is most prevalent in mathematics.

There are different forms of this stratification. The most common is when students are assessed in mathematics and then grouped according to the results on that assessment. While elsewhere different terms are used, in Australia this is termed streaming. There are also many schools which select one or more high-achieving groups, but otherwise have the rest of the groups grouped heterogeneously (commonly described as tracking).

There is limited research on the extent of streaming practices in Victorian schools. In a detailed study, using an on-line survey of grouping practices in mathematics, Forgasz (2010) reports that 80 % of the 44 responding schools had some form of streaming in the years 7–10, with three quarters of the respondents indicating support for that streaming. Indeed, of the four schools which said that there was no streaming, three of the respondents were opposed to that policy. It can be inferred that the teachers who responded were overwhelmingly in favour of some streaming. A positive characteristic of the responses was that only 37 % of the schools reported streaming in year 7, 55 % in year 8 and 70 % in year 9. This indicates that the majority of these schools have heterogeneous groupings at year 7 and half of the schools at year 8. Clearly, it is possible to teach mathematics without streaming.

At the same time, it is easy to understand attempts to make grouping more homogeneous. Some of the difficulties that have been identified with heterogeneous grouping are:

- Teachers set expectations and starting points based on low-achieving students and as a result the other students are under-extended and less satisfied with their learning environment.
- Teachers over direct the learning of everyone (assuming low-achieving students cannot cope), which has the effect of encouraging a fixed mind set (Dweck 2000) and a passive approach to learning in the students.
- There is negative peer pressure on hard-working students, which is very real. Sullivan et al. (2006), for example, found that the classroom culture exerts a more powerful (negative) influence on students than their individual aspirations.
- Teachers can ignore the diversity of readiness and instead treat everyone as the same (possibly by giving routine tasks that everyone can and is willing to do) (see Doyle 1986).

- Teachers teach different content to different groups, which not only increases the teachers' workload, but destroys any sense of a classroom community.
- Low-achieving students 'performance avoid' (Elliot 1999) by misbehaving, being a group-work passenger or pretending to work while not actually doing anything.

Clearly, if heterogeneous groups are to maximise the learning of all students substantial actions must be taken by teachers and schools to address these difficulties.

On the other hand, homogeneous groupings can have the effect of restricting student opportunities if:

- Teachers teach different content to different groups, thereby not only narrowing the options of some students but also actually closing them off too early.
- There is limited or no movement between groups, which appears to be the most common situation. If there is no chance of 'promotion', students are unlikely to try hard and, in any case, students develop affiliation with the group and so do not want to move.
- Steps are not taken to avoid development of poor self-concept by some members of the upper streams. This was described by Marsh et al. (2005) as the Big fish little pond effect in which a substantial minority of students in such groups develops a low self-concept and subsequent little interest in the subject because they feel that their classmates are more able than they are.

There are, however, significant barriers to overcoming negative effects of homogeneous grouping. It is very difficult to ensure that students in all groups have the same opportunities if the curriculum is stratified and only a limited subset of the curriculum is offered to some groups. This can be exacerbated if teachers feel that skills precede other learning and so emphasise skills to the detriment of other aspects of mathematics, such as communication, meaning and relevance. Indeed, the very placement of students in low streams communicates to students that their teachers think they cannot learn. A further risk is that the 'homogeneous' grouping of students communicates to teachers that the students are indeed of like achievement.

In terms of seeking advice from research, Forgasz (2010) reviewed a range of studies on the effects of grouping classes of students by their achievement. She argues that the results are

> inconclusive, particularly for those at the highest levels of achievement. There is general agreement, however, that those in middle and lower achieving mathematics classes may be disadvantaged with respect to achievement, and that their future mathematics and life options are likely to be curtailed. (p. 66)

More conclusively, in a major metanalysis, Hattie (2009) argues that stratification, streaming, tracking, setting has 'minimal effect on learning outcomes and profound negative equity effects' (p. 90). He argued that low-stream classes are 'deadening, non educational environments' (p. 90) that fail 'to foster the outcomes schools value' but are focused on 'remediation through dull, repetitious seatwork'. Yet Hattie also argued that 'it seems that the quality of teaching and the nature of students' interactions are the key issues, rather than the compositional structure of the classes' (p. 91).

In other words, streaming students for mathematics poses a threat to equity and opportunity but it is as much the ways classes are taught that is important as it is the method of grouping.

Self-Fulfilling Prophesy Effects

It is clear that both homogeneous and heterogeneous grouping practices create challenges for teachers and schools. To explore further the challenges that teachers experience, a particular effect that applies to both forms of grouping is outlined. It seems that if teachers believe that students are less likely to learn mathematics, then those students have restricted opportunities to learn.

This effect, referred to as *self-fulfilling prophesy,* has been reviewed over a broad range of contexts and the findings suggest that the effect applies when teachers attribute particular characteristics to students, such as whether they are high achieving or likely to experience difficulty in learning. The effect has also been noted in terms of teachers forming judgments about student potential based on race or ethnicity. For example, in making recommendations on ways to repay what he describes as an 'educational debt' to Maori students in New Zealand, Bishop (2010) refers to the 'dominance of a deficit discourse among teachers' (p. 130). He identifies this as the single pedagogical issue that needs to be addressed in teaching Maori students as it continues to be a major barrier to educational reform. Bishop calls for action to address the situation that 'currently the majority of teachers are defining Maori potential in deficit terms' (p. 134). This effect operates similarly in many classrooms and schools, and has a negative impact on the learning opportunities of students whom the teachers see as low achieving.

Almost three decades ago, Brophy (1983) posed a cyclic model that describes how this self-fulfilling prophecy might operate:

Step 1: Teachers form early differential expectations for students.

Step 2: As a result, the teachers behave differently to different students and this differential behaviour communicates the teachers' expectations to the students. If such treatment of the students is consistent, and if the students do not resist, it will have an effect on their self-concept, achievement, motivation, aspirations and classroom conduct.

Step 3: The students' responses will actually reinforce the teacher's original expectations. Ultimately there will be differences in student achievement and outcomes directly due to this effect.

In other words, the effect in schools and classes is connected to the responses that teachers give to different types of students. Brophy (1983) argued that, for those students considered to have difficulty in learning mathematics, teachers:

- wait less time for them to answer questions;
- give them the answer or call on someone else rather than waiting;
- use inappropriate reinforcements;
- criticise them more for failure and praise them less frequently;
- do not give them public feedback on their responses;
- call on them less to respond;
- demand less; and
- have less friendly verbal and nonverbal contact.

One of the explanations for this effect relates to what Brophy called the teachers' *need for control*. For example, when dealing with students whom the teachers think can learn well, teachers feel more able to predict student behaviour when interacting with them both privately and publicly, no matter who initiates the interaction. On the other hand, if teachers are worried about classroom control they are likely to avoid public interactions with low-expectation students, especially interactions the students have initiated. Teachers may call on students less if they believe that they will experience difficulty learning, and ignore or discount their attempts to initiate questions.

Another explanation for the way the effect works is related to *attribution*. In this, teachers who attribute to themselves a student's failure to grasp an idea are likely to give further explanations and to seek other ways of explaining the difficult idea. If however teachers attribute the failure to a student's lack of ability or some other characteristic, they may give up and move the student on to some other simpler task, thereby reducing the likelihood that the student will learn the intended content.

A third explanation is related to the notion of *learning through challenge*. Sullivan et al. (2011) argue that students are more likely to learn mathematics when they work on problems that they cannot yet do, as distinct from only practising routines that they already know. If teachers are not presenting low-achieving students with challenges, this also reduces their opportunities to learn.

Interestingly, Brophy (1983) argued that being aware of the potential impact self-fulfilling prophesy effects on particular groupings can minimise negative effects. For example, teachers in both heterogeneous and homogeneous groups can avoid the impact of their presumptions about the potential of individual students by consciously treating all students similarly. Further, even class groups designed to maximise homogeneity are diverse in their readiness. The explanation for this is that even though students are grouped by their achievement, the intent of streaming is to group students by their ability, which cannot be measured directly. Because there are high-ability students who may achieve low scores on a particular assessment for a variety of reasons, and some low-ability students who score well, perhaps due to their effort or even out of school tuition, even like achievement groups will have a range of abilities. In other words, if the students are grouping by their achievement, there will still be a diversity of readiness to learn mathematics and all teachers need to plan to address this diversity.

Table 14.1 Comparison of students' 'confidence' and 'satisfaction' across the year levels

Year level	*N*	Confidence mean	Confidence SD	Satisfaction mean	Satisfaction SD
Year 5	302	4.70	1.38	4.40	1.65
Year 6	392	4.95	1.24	4.28	1.52
Year 7	126	4.61	1.33	3.82	1.55
Year 8	110	4.91	1.25	4.15	1.76
All	930	4.81	1.31	4.24	1.60

SD standard deviation

Further Data on the Diversity in Classrooms

The following explores this issue of within-class diversity further. Sullivan et al. (2013) present findings that illustrate the issue of diversity and the need to address it within each classroom. They administered a survey to students over 17 schools and 95 classes designed to capture some attitudes to mathematics. The first items in the survey asked the students to indicate on a scale of 1–7, 'How good are you at maths?' and 'How happy are you in maths class?' In the following discussion, responses to the first item are taken to be a measure of confidence, and responses to the second to be a measure of satisfaction.

The students' responses were predominantly positive although they did vary from 1 to 7 indicating that there was a range of levels of confidence and satisfaction in every class. Interestingly, there was not much difference between the responses of the students at the respective year levels, given that junior secondary students appear to be less satisfied and less confident in their ability. Table 14.1, reproduced from Sullivan et al. (2013), presents a comparison of responses of students across these year levels.

Overall, the students seemed to feel more confident than satisfied and there is a substantial spread of scores at each of the grade levels, meaning that while there were some students who gave positive responses, there were others who gave negative ones. While there were statistically significant differences between the grade levels for both confidence ($F(3926)=3.34$, $p<0.02$) and satisfaction ($F(3926)=4.11$, $p<0.01$), the differences within each year level were more substantial than those between year levels. In other words, the differences within year levels are more substantial than those between year levels and rather than teaching differently between years, the implication is that teachers need to address differences in confidence and satisfaction in the classes they are teaching whatever the level and whatever the grouping practice.

From another perspective on the challenge of addressing diversity, the following data are from a current project, with David Clarke and Doug Clarke, in which we surveyed a group of primary and secondary teachers working with us on the implementation of the AC. Table 14.2 presents the responses of teachers to a survey on classroom grouping practices. Nineteen teachers were asked to indicate the percentage of their lessons in which they used each of the nominated practices.

Table 14.2 Responses of teachers reporting frequency of use of particular grouping practices ($n=19$)

Method of grouping	Mean % lessons
Students grouped by achievement with each group working on related tasks	30
Students grouped by achievement with each group working on unrelated tasks	4
Whole-class teaching, with everyone working on the same tasks with you assisting individual students, with some students completing more than others	27
Whole-class teaching, working on similar tasks, differentiated for students who experience difficulty in starting and/or who are ready for more challenge	28
Whole-class teaching, with everyone working on similar tasks and you choose a like achievement group with whom you work for most of the lesson	4

It seems that nearly all lessons involved students working on related or similar tasks, around one-third of lessons have students grouped by achievement, one-third of lessons have students completing a different quantity of tasks and around one-third of the lessons involved students working on differentiated tasks. These findings affirm that teachers are identifying ways to address the differences in readiness.

A Model of Teaching to Address Differences in the Preparedness of the Students

A particular model of teaching, designed to address the diversity in student readiness in mathematics (whatever the method of classroom grouping), was proposed by Sullivan et al. (2009a). It has five elements.

The first element relates to building a *communal classroom experience*. Sullivan et al. (2009b) argue that all students should have at least some core experiences that can form the basis of later discussions. The expectation is that teachers work with students to develop in them a sense of membership of the class as a whole. This notion is based on Wood's (2002) research which emphasises how 'social interactions with others substantially contribute to children's opportunities for learning' (p. 61) as well as the interplay between children's developing cognition and the 'unfolding structure that underlies mathematics' (p. 61). Integral to this element is the assumption that mathematical communications in classrooms that are intended to include all students can best occur if there is some communal experience. If some students in a class are excluded from common experiences and are unable to participate in discussion, this voids the possibility of them feeling affiliated with the class as a whole. Further, such experiences not only create opportunities for social interaction but also promote thinking about mathematics.

The second element is the planning of a *trajectory* of mathematical tasks. Sullivan et al. (2009b) argue that there are two considerations for the trajectory of tasks. The first is that there are benefits to inclusivity if at least some of the tasks are open ended. A number of researchers have argued that open-ended tasks engage students in thinking about mathematics exploration, enhance motivation through increasing sense of control and encourage students to investigate, make decisions, generalise, seek patterns and connections, communicate, discuss and identify alternatives (Christiansen and Walther 1986; Middleton 1995; Sullivan 1999). The second consideration is that earlier tasks in the sequence provide experiences that scaffold students in the solution of later tasks, allowing them to engage in more sophisticated mathematics than would otherwise have been the case. This connects directly with the notion of a hypothetical trajectory.

There are different ways to create sequences of tasks. One type of sequence is where the problem formulation remains constant but the numbers used increase the complexity of the task, say moving from small numbers to larger numbers. Another type of sequence is where the problem is progressively made more complex by the addition of supplementary steps or variables, such as in a network task where additional nodes are added. A third type of sequence may be where the concept itself becomes more complex, such as in a sequence of finding areas or progressively more complex shapes from rectangles, to composite shapes, to irregular shapes. The creation of such sequences is a key component of the planning model.

The third element involves *enabling prompts* that are posed to engage students experiencing difficulty. Students are more likely to feel fully part of the class if teachers offer prompts to allow those experiencing difficulty to engage in active experiences related to the initial goal task, rather than, for example, requiring such students to listen to additional explanations or assuming that they will pursue goals substantially different from those of the rest of the class. There are some generic types of prompts. For example, it nearly always helps to draw a diagram or model, to remove one of the constraints, to offer more choice, or to change the form of representation.

A fourth element relates to anticipating that some students may complete the planned tasks quickly, and can be posed supplementary tasks that *extend their thinking* on that task. One of the characteristics of open-ended tasks is that they create opportunities for extension of mathematical thinking, since students can explore a range of options as well as considering forms of generalised response. In practice it is arguable that this is the most important and challenging of these planning elements. The premise is that the students in the class progress together through the lesson contributing to the sense of communal experience. Unless creative opportunities are provided for the students who have completed the tasks along the way then not only might they be bored, and so create difficulties for the teacher, but also they will not be using their time effectively. Note that this offers substantial advantages over the strategy of moving students who finish the work onto the next chapter of the text.

The fifth element of the framework is related to being explicit about the pedagogies of mathematics teaching. This is derived from the work of Bernstein (1996) who described pedagogies that are hidden from some students. Bernstein argued that, through different methods of teaching, students receive different messages about the overt and the hidden curriculum of schools. He suggested that some students are able to make sense of this 'invisible' pedagogy more effectively than others, due to their familiarity with the embedded sociocultural norms, and hence those students have more chance of success. As suggested by Delpit (1988), Zevenbergen

(1998) and Dweck (2000), it may be possible to moderate the effect of the hidden curriculum by explicit attention to aspects of pedagogies associated with such teaching. Sullivan et al. (2002) list a range of strategies that teachers could use to make implicit pedagogies more explicit and so address aspects of possible disadvantage of particular groups. It seems that teachers are able to make explicit at least some of the key pedagogies associated with such teaching, and that students respond to this explicitness in the direction intended.

In other words, it is proposed that better learning outcomes are likely if lessons are based on sequenced tasks that have potential for students' decision making, in which it is intended that all students participate, with prompts for those students who are experiencing difficulty and those who complete the work quickly, and for which the desired pedagogies are made explicit.

An Illustrative Example of Such Teaching

To exemplify what such teaching might look like, the following plan is presented to illustrate the elements of a lesson based on this model. The plan is written for an actual lesson hypothetically to be taught by other teachers. It is stressed that the project cited above gathered evidence of the applicability and effectiveness of this approach.

Introduction

Explain the mathematical focus of the lesson which is on ways of representing data sets using single scores, especially mean, median, mode and range.

Outline the expectation that students:

- show how they got their answers on the worksheets,
- give more than one possible solution strategy,
- keep trying even if it is difficult (it is meant to be),
- explain your thinking, and
- listen to other explanations.

Task 1: Writing a Sentence

Pose the following problem:

- Write a sentence with 5 words, with 4 as the mean of the number of letters in the words. Do not use any words of 4 letters.

Invite questions for clarity, but will not tell them how to do it. The students can get started without further explanation.

Possible enabling prompts are:

- Write a sentence of 5 words.
- Write a sentence of 20 letters.

As an extending prompt, ask students to make one of the words as long as possible.

Task 2: Fishing

Pose the following problem:

Seven people went fishing. The mean number of fish caught was 5, the median was 4 and the mode was 3. How many fish might each of the people have caught? (Give at least three answers.)

Possible enabling prompts are:

- Seven people went fishing. The mode of the number of fish caught was 3. How many fish might each of the people have caught?
- Seven people went fishing. The median number of fish caught was 4. How many fish might the each of the people have caught?

Possible extending prompts are:

- Find all the possible answers if the range is 6.
- What if it was 6 people that went fishing? What different does that make?
- What is the maximum number of fish that an individual person might have caught?

In summary, the lesson exemplifies the planning model in which it is designed to maximise the chances that the whole class can feel part of the learning community, there is a trajectory of challenging tasks, tasks have both enabling and extending prompts, and there is some attempt to be explicit about the pedagogies.

Conclusion

The (Australian) curriculum is written assuming schools and classrooms will be structured to maximise learning opportunity. This chapter has described specific challenges facing Australian schools and, in particular, the large within-school differences in achievement. It has argued that even though there are challenges associated with teaching both heterogeneous and homogeneous groups, it seems that homogeneous groupings have the potential to restrict the opportunity to learn of some students. It was also argued that it is not so much the method of grouping but the approach that teachers take to addressing diversity that ensures that the needs of all students are addressed. A teaching model was presented which is designed to address the diversity of preparedness, whatever the classroom grouping.

References

ACARA. (2009). Shape of the Australian curriculum: Mathematics. Downloaded in Feb 2012 from http://www.acara.edu.au/verve/_resources/Australian_Curriculum_-_Maths.pdf. Accessed Sept 2013.

ACARA. (2012). The shape of the Australian curriculum. http://www.acara.edu.au/verve/_resources/The_Shape_of_the_Australian_Curriculum_V3.pdf. Accessed Jan 2012.

Bernstein, B. (1996). *Pedagogy, symbolic control, and identity: Theory, research, critique*. London: Taylor & Francis.

Bishop, R. (2010). Closing the gap in education by addressing the educational debt in New Zealand. In I. Snyder & J. Nieuwenhuysen (Eds.), *Closing the gap in education? Improving outcomes in Southern World societies* (pp. 129–148). Australia: Monash University.

Brophy, J. E. (1983). Research on the self-fulfilling prophecy and teacher expectations. *Journal of Educational Psychology, 75*(5), 631–661.

Christiansen, B., & Walther, G. (1986). Task and activity. In B. Christiansen, A. G. Howson, & M. Otte (Eds.), *Perspectives on Mathematics education* (pp. 243–307). Dordretcht: Reidel.

Delpit, L. (1988). The silenced dialogue: Power and pedagogy in educating other people's children. *Harvard Educational Review, 58*(3), 280–298.

Doyle, W. (1986). Classroom organisation and management. In M. C. Wittrock (Ed.), *Handbook of research on teaching* (pp. 392–431). New York: Macmillan.

Dweck, C. S. (2000). *Self theories: Their role in motivation, personality, and development*. Philadelphia: Psychology Press.

Elliot, A. J. (1999). Approach and avoidance motivation and achievement goals. *Educational Psychologist, 34*(3), 169–189.

Forgasz, H. (2010). Streaming for mathematics in years 7–10 in Victoria: An issue of equity? *Mathematics Education Research Journal, 22*(1), 57–90.

Hattie, J. (2009). *Visible learning: A synthesis of over 800 meta analyses relating to achievement*. New York: Routledge.

Marks, G., McMillan, J., & Hillman, K. (2001). Tertiary entrance performance: The role of student background and school factors. http://research.acer.edu.au/lsay_researh/24/ Accessed Sept 2013.

Marsh, H. W., Craven, R. G., & McInerney, D. (Eds.). (2005). *New frontiers in SELF research*. Greenwich: Information Age Press.

McGaw, B. (2007). *Internationalizing conceptions of quality in education*. Presidential session symposium, American education research association conference, Chicago, April.

Middleton, J. A. (1995). A study of intrinsic motivation in the mathematics classroom: A personal construct approach. *Journal for Research in Mathematics Education, 26*(3), 254–279.

Rothman, S., & McMillan, J. (2003). Influences on achievement in literacy and numeracy. http://www.acer.edu.au/documents/LSAY_lsay36.pdf. Accessed Sept 2013.

Sullivan, P. (1999). Seeking a rationale for particular classroom tasks and activities. In J. M. Truran & K. N. Truran (Eds.) *Making the difference*. (Proceedings of the 21st conference of the mathematics educational research group of Australasia, pp. 15–29). Adelaide.

Sullivan, P., Tobias, S., & McDonough, A. (2006). Perhaps the decision of some students not to engage in learning mathematics in school is deliberate. *Educational Studies in Mathematics, 62,* 81–99.

Sullivan, P., Mousley, J., & Jorgensen, R. (2009a). Tasks and pedagogies that facilitate mathematical problem solving. In B. Kaur (Ed.), *Mathematical problem solving* (pp. 17–42). Association of Mathematics Educators: Singapore/USA/UK World Scientific Publishing.

Sullivan, P., Prain, V., Campbell, C., Deed, C., Drane, S., Faulkner, M., McDonough, A., Mornane, A., & Smith, C. (2009b). Trying in the middle years: Students' perceptions of their aspirations and influences on their efforts. *Australian Journal of Education, 5*(2), 176–191.

Sullivan, P., Cheeseman, J., Michels, D., Mornane, A., Clarke, D., Roche, A., & Middleton, J. (2011). Challenging mathematics tasks: What they are and how to use them. In L. Bragg (Ed.), *Maths is multi-dimensional* (pp. 33–46). Melbourne: Mathematical Association of Victoria.

Sullivan, P., Clarke, D., & Clarke, B. (2013, in press). *Teaching with tasks for effective mathematics learning*. New York: Springer. http://www.springer.com/education+%26+language/mathematics+education/book/978-1-4614-4680-4. Accessed Sept 2013.

Sullivan, P., Zevenbergen, R., & Mousley, J. (2002). Contexts in mathematics teaching: Snakes or ladders? In B. Barton, K.C. Irwin, M. Pfannkuch & M. Thomas (Eds), *Mathematics Education in the South Pacific: Proceedings of the 25th annual conference of the Mathematics Education Research Group of Australasia* (pp. 649-656), Mathematics Education Research Group of Australasia, Auckland, July.

Thompson, S., De Bortoli, L., Nicholas, M., Hillman, K., & Buckley, S. (2010). *Challenges for Australian education: Results from PISA 2009*. Melbourne: Australian Council of Educational Research.

Wood, T. (2002). What does it mean to teach mathematics differently? In B. Barton, K. C. Irwin, M. Pfannkuch, & M. Thomas (Eds.), *Mathematics education in the South Pacific* (pp. 61–71). Auckland: Mathematics Education Research Group of Australasia.

Zevenbergen, R. (1998). Language, mathematics and social disadvantage: A Bourdieuian analysis of cultural capital in mathematics education. In C. Kanes, M. Goos, & E. Warren (Eds.), *Teaching mathematics in new times* (pp. 716–722). Gold Coast: MERGA.

Chapter 15
Commentary for Section 3: From Diversity to Practices: Addressing, Redressing and Taking Action

Laurinda Brown

In Peter Sullivan's opening chapter, he sets out the intentions of this book and also presents me with a perspective through which to discuss issues arising after reading the three chapters in this section. In the introduction to Chapter 1 (p. 3), Peter asks, "whether the goal of any recommendations for change is to improve the education of all students, without addressing the differences, or to find ways to reduce the differences between groups of students". In the conclusion to the chapter, as researchers addressing inclusivity, we are asked to report on "what redressing disadvantage might look like" (p. 13). I will comment on each chapter in these terms, particularly highlighting what any advice for implementation might be, in order to discuss, from a UK perspective, different levels of advice and their relation to actions.

There are many audiences for any advice that researchers might wish to see implemented and, in these comments, I am particularly interested in what individual teachers or communities of teachers in a particular school might be able to do with the advice offered.

Firstly, I will use two personal stories to provoke a need for the link to the writing in the chapters in this section.

When I went to university to read mathematics I volunteered to read for a blind student from the USA. He was a historian. He was independent as he moved around the city and was a familiar figure as his walking cane swept before him as he sped along.

In the UK, at that time, there were special schools for blind children, not so in the USA. I was not used to seeing blind people in the UK being so independent. I began to question the practices of special education that excluded students from ordinary school. I had a speech impediment as a child, saying "w" for "r", so "twain" for "train". Putting me in a school for students with speech impediments seems ludicrous, why was being blind any different?

L. Brown (✉)
Graduate School of Education, University of Bristol, Bristol, UK
e-mail: laurinda.brown@bristol.ac.uk

A. Bishop et al. (eds.), *Diversity in Mathematics Education*,
Mathematics Education Library, DOI 10.1007/978-3-319-05978-5_15,

The focus on teachers' actions is in response to the move away from students with special needs being "integrated" into mainstream education (the students had to change and conform) to inclusion, where the teacher is expected to be aware of the needs of each of their students and adapt the classroom environment and curriculum so that each child makes progress. For an individual teacher this may seem like rhetoric, so finding ways for teachers to see things they can try out or "do" to create an inclusive learning culture in their classroom needs to become a central tenet for policy makers and researchers.

I went to a girls' grammar school (about 20 % of the population in those days), which was streamed in the UK sense of the word, that is, results for all subjects were averaged out and the top stream contained the students with the highest scores. We were then taught together as a class for all our subjects. I could do mathematics, delivered as exposition and practice, but many of my high-achieving classmates could not. Many girls in that top stream failed their final mathematics examination (aged 16). I felt passionately that there had to be a different way to teach mathematics than this.

In the language of this book, for the girls who failed their need to make mathematics mean something was not being addressed. These girls were expected to adapt themselves to learn mathematics in a particular way that they found meaningless and were consequently not engaged (see Chapter 12). There was no link to real-life or even imaginable mathematics in the sense of the Freudenthal Institute's Realistic Mathematics Education. Exposition and practice, of similar examples treating the class as a whole, was not inclusive and did not "develop students' confidence in their ability to do mathematics" (p. 198).

I am now working with prospective teachers on a 1-year post-graduate course (PGCE) in the UK. Many of my student teachers have had positive experiences of learning mathematics in their own schooling, largely through exposition and practice. They see learning as being successful at rehearsing the algorithm that had been explained to a class of children, each of whom are similarly rehearsing and all being successful. They do not all, therefore, at the start of the year, feel the need that I did that mathematics had to be taught differently. The first task for them is to open up to different possibilities for their actions as teachers. This can be from seeing someone teaching in a way they feel able to do, or watching a video (although this is sometimes problematic, see Coles 2014) and always from being aware of the range of experiences of others, most powerfully, a prospective teacher who had himself been taught in a mixed ability class from age 11 to 16. When others in the PGCE group expressed a view that mixed ability teaching was wrong or difficult to do, he would speak eloquently of the strengths: you knew everyone and shared all lessons in a day where you realized that everyone had talents and subjects where they needed help. The classroom was a community where the students worked together and supported each other in whatever activities were on offer from the teachers.

The teaching and learning environment is complex and every action of a teacher cannot be specified down to the last detail. As professionals, teachers make decisions and so one question for a researcher who wants to impact learning is what can be said or done in sufficient detail that there is motivation to act while leaving enough space for the skills of an individual teacher to be able to be used? In Chapter 1, Sullivan illustrates "challenges of offering succinct advice" (p. 8) with research findings in school leadership written as benchmarks that could be used to evaluate schools. These statements are not given as actions for teachers to "do". However, one benchmark is that "Explicit and clear school-wide targets for improvement have been set …" (p. 8). When faced with statements such as these I ask, what do you want me to do? And if I cannot answer that question, then the statements for me are at a different level to that of the action that this book seeks to be about. There is a call to action—has this been done? The details of what the targets are, how to negotiate them (if at all) and how to ensure the implementation of the targets across the institution uniformly to achieve impact are not given. Here there is a large space for management to work out a strategy for inclusion in the school. No activities as such are provided.

In "Maximising opportunities in mathematics for all students: Addressing within school and within class differences", which is the title of Sullivan's Chapter 14, in this section, he gives a "model of teaching to address differences in the preparedness of students" (p. 248), developed from the research literature, through which lessons can be planned to be more inclusive, with some principles for action:

- That all students should have at least some core experiences that can form the basis of later discussion;
- Planning a trajectory of mathematical open-ended tasks where the mathematical concept becomes more complex and there is scaffolding for students in the solution of later tasks;
- Enabling prompts (e.g., draw a diagram) are given;
- Extending of tasks is planned for;
- Being explicit about the pedagogies of mathematics teaching. (distilled from pp. 8–10)

These principles aim to support "better learning outcomes" through "sequenced tasks that have potential for students' decision making" (p. 250) and are closer to actions, with space for choices and decisions to be made by the individual teacher or structuring planning discussions in a department. Discussion of this model is followed by an illustrative example lesson plan (p. 250) where each aspect is considered and a trajectory of activities given.

This framework will become one of many that I offer my prospective teachers on their journeys to find the teacher they can become, so the model has already had some impact. The children in classrooms, also, have to find the mathematics learner they can be. The link to taking action seems to be through space for decision-making, and, from reading these chapters, setting up a culture in the classroom that works with student engagement that also pays attention to the wider context outside the school and pays attention to feelings. What do I do as a teacher to set up a classroom in which it is alright to be wrong?; where we learn from our mistakes through challenge?; where I, as the teacher, am learning the children, not labelling them but paying attention to their thoughts, feelings and mathematical actions, adapting myself contingently to the needs of the individuals while using the wider community?

As an alternative approach, Cotton, discussing research he did in 2010 (Coles et al. 2013), introduces aims to meet the following requirements of anti-racist educational processes:

- Develop learners' understandings of cultures other than their own;
- Enable learners to reflect on and develop positive attitudes towards cultural and linguistic diversity;
- Use resources which draw on learners' cultural heritage and experience and which counter or challenge bias;
- Use familiar contexts as starting points;
- Illustrate the diverse cultural heritage of mathematics;
- Critique and challenge stereotypical views of particular groups of people through the analysis of data;
- Encourage collaborative learning (pp. 95–96).

These statements have overlaps with Sullivan's criteria and could also become benchmarks for a school to use to evaluate its teaching. Again, the "process was supported by offering teachers a planning *pro forma* to bring the ideas above into preparing lessons" (p. 96), under the headings, context of lesson, grouping of learners, resources, language, mathematics and collaborative learning. In each section, questions are provided to support teachers in their planning, for instance, for the heading "collaborative learning", suggested questions are "To what extent does the lesson:

- Encourage learners to express and examine their own views?
- Encourage learners to become involved in their own learning?
- Encourage learners to pose their own problems? (p. 97)

Both these frameworks avoid the trap of being too specific, such as having a script to follow, but there still might not be enough support for some teachers without the sort of exemplar introduced by Sullivan or through observing a teacher practising using the policy or watching videos of practice. I have found that lesson write-ups, detailing the beginning sequence of a lesson followed by possible prompts for further ways in which the project might develop and sharing possible metacomments that the teachers can make related to the potential mathematical behaviours of the children (e.g., getting organized, specializing and generalizing, getting something wrong supports you learning something) support the developing culture in a classroom. Guidelines that address the issues need to support teachers taking action and yet leave choice for the teachers to redress imbalances in their own contexts.

For the National Strategy in the UK, strategies for teachers to use for children to learn number, including addition of single digits (Chapter 13), were distributed (Department for Education 2010). Some of this advice, at a detailed level, was developed from Anghileri's research that documented in detail the different strategies used by children in adding two single-digit numbers. She states that "[w]atching children will help teachers assess the strategies they are using [such as] using an addition strategy (counting from 7 to 9) but others will need this idea to be pointed out explicitly" (2006, p. 57) This way of learning the children as part of teaching seems important for inclusion and I was not so sure that the patterns of difficulty (outlined in Chapter 13, p. 233) might not become a form of labelling? What the patterns of difficulty do add is another dimension to the awareness of addition strategies that each teacher needs. However, if "inefficient counting" is identified, the teacher has to have thought through or been shown, what might be offered as activities to the child to support them in developing.

So, there are various levels of advice aimed at particular populations. What is important to realize is that there is always a need for further work following the advice. This is clear in the case of the benchmarks specified above, but less clear perhaps with more detailed advice aimed at individual teachers, like schemes of work within a department that might include detailed lesson plans or central advice from governments. The advice given as frameworks for action or questions to consider while planning, alongside exemplars, videos or lesson write-ups of the sorts of lessons that have led to learning seem useful ways that teachers can gain awareness of what new practices are like so that they can plan with conviction to act in their classrooms inclusively.

References

Anghileri, J. (2006). *Teaching number sense*. London: Continuum.

Coles, A. (2014). Mathematics teachers learning with video: the role, for the didactician, of a heightened listening. *ZDM - The International Journal on Mathematics Education, 46*(2), 267–278.

Coles, A., Barwell, R., Cotton, T., Winter, J., & Brown, L. (2013). *Teaching secondary mathematics as if the planet matters*. London: Routledge.

Department for Education. (2010). Teaching children to calculate mentally. http://www.ness.uk.com/maths/Guidance%20Documents/Teaching%20children%20to%20calculate%20mentally.pdf. Accessed 9 Nov 2013.

Index

A. Bishop et al. (eds.), *Diversity in Mathematics Education*,
Mathematics Education Library, DOI 10.1007/978-3-319-05978-5,

The manufacturer's authorised representative in the EU is Springer Nature Customer Service Centre GmbH, Europaplatz 3, 69115 Heidelberg, Germany. If you have any concerns regarding our products, please contact ProductSafety@springernature.com

Printed and bound by CPI Group (UK) Ltd, Croydon, CR0 4YY

22/09/2026

02221069-0002